METHODS IN MOLECULAR BIOLOGY

For further volumes:
http://www.springer.com/series/7651

For over 35 years, biological scientists have come to rely on the research protocols and methodologies in the critically acclaimed *Methods in Molecular Biology* series. The series was the first to introduce the step-by-step protocols approach that has become the standard in all biomedical protocol publishing. Each protocol is provided in readily-reproducible step-by-step fashion, opening with an introductory overview, a list of the materials and reagents needed to complete the experiment, and followed by a detailed procedure that is supported with a helpful notes section offering tips and tricks of the trade as well as troubleshooting advice. These hallmark features were introduced by series editor Dr. John Walker and constitute the key ingredient in each and every volume of the *Methods in Molecular Biology* series. Tested and trusted, comprehensive and reliable, all protocols from the series are indexed in PubMed.

Sepsis

Methods and Protocols

Edited by

Wendy E. Walker

Department of Molecular and Translational Medicine, Texas Tech University Health Sciences Center at El Paso, El Paso, TX, USA

Humana Press

Editor
Wendy E. Walker
Department of Molecular and
Translational Medicine
Texas Tech University Health Sciences
Center at El Paso
El Paso, TX, USA

ISSN 1064-3745 ISSN 1940-6029 (electronic)
Methods in Molecular Biology
ISBN 978-1-0716-1487-7 ISBN 978-1-0716-1488-4 (eBook)
https://doi.org/10.1007/978-1-0716-1488-4

Cover Illustration Caption: See Chapter 9, Figure 2 of this volume for more details.

This Humana imprint is published by the registered company Springer Science+Business Media, LLC, part of Springer Nature.
The registered company address is: 1 New York Plaza, New York, NY 10004, U.S.A.

Preface

Sepsis is a condition that occurs when the immune response to an infection becomes dysregulated, resulting in life-threatening organ dysfunction [1]. This immune dysregulation is comprised of overwhelming and persistent inflammation, paired with immune suppression [2–5]. To survive, a patient must defeat the infection and return to immune homeostasis. Sadly, many patients do not achieve a favorable outcome. Sepsis is the leading cause of in-hospital death [6]. Furthermore, up to 50% of patients who survive sepsis experience long-term physical and psychological effects, and many experience repeated hospitalizations [7]. Sepsis is especially devastating for people who are elderly, or immunosuppressed; these individuals experience higher rates of sepsis, higher in-hospital mortality, and a poor long-term prognosis [8]. Sepsis is also prevalent in the intensive care unit (ICU) [9]. Trauma and the use of mechanical ventilators increase the risk of nosocomial infections and, given that most ICU patients have pre-existing inflammation and impaired defense, these can quickly develop into sepsis. Recently, coronavirus infectious disease 2019 (COVID-19) has emerged as a major global pandemic, caused by the SARS-CoV-2 virus. This virus has taken an enormous toll on human lives and health. It is clear that many patients who become critically ill with COVID-19 develop features of sepsis, and secondary bacterial infections are prevalent. More than ever before, sepsis is a global health priority.

A great deal of effort has been devoted to improving the identification and treatment of septic patients. At the date of this publication, the Sepsis-3 guidelines are the current protocol to identify patients at risk of death from sepsis [1]. Organ dysfunction is identified via the sequential organ failure assessment (SOFA) score (based on measurements of arterial blood gas data, platelets, bilirubin, creatinine and urine output, as well as blood pressure and the Glasgow coma score). The quick SOFA (qSOFA) score was also developed to rapidly screen for sepsis mortality risk (based on respiratory rate, blood pressure, and mental status) and can be rapidly completed at the bedside and in settings with limited resources. Once a patient is identified to be at risk of sepsis, specific treatment regimens (bundles) should be administered within a defined time frame [10]. Treatments for sepsis include antibiotics, fluids, and vasopressors. Although these treatments are basic, their early implementation greatly improves the survival of patients with bacterial sepsis. Ideally, blood and other bodily fluids are obtained before the first antibiotic is administered and are cultured to test for the presence of an infection and to identify the microbial agent. This allows fine-tuning of the antibiotic regimen. However, blood culture can take up to 48 h to yield results, and many patients with a real infection remain culture negative. Despite carefully managed treatment, the mortality rate remains high in patients with severe sepsis and septic shock (30–50%) [11]. Therefore, it is imperative to develop new treatments for sepsis.

Sepsis remains a conundrum, as it is difficult to achieve a definitive early diagnosis (before organ dysfunction occurs), yet early treatment is critical to avoid a lethal outcome. Because of this, clinicians must err on the side of caution, administering antibiotics and retaining patients in the hospital until the blood culture results and other clinical tests are returned. This increases the potential for antibiotic-resistant bacteria to develop in the hospital setting and may delay appropriate treatment for patients with other conditions that mimic sepsis. Therefore, it is imperative that we develop improved tests that achieve rapid, sensitive, and specific sepsis diagnosis.

The pathophysiology of sepsis is quite complex and involves multiple organ systems. *In vitro* models alone are not adequate to study this condition. The basic biology of mammals is highly conserved between species. Mice are most often chosen for sepsis research, because they are the lowest sentient species that can adequately mimic the disease process. Furthermore, genetically modified mouse strains are useful for mechanistic studies. Rats are also commonly used for sepsis research. The signs of sepsis in rodents closely resemble those in humans. Recently, the use of mice in sepsis research has received some debate. Many therapies that were developed in mice did not yield positive results in clinical trials. Moreover, conflicting reports were published regarding whether the genomic response to inflammation is similar between mice and humans [12, 13]. However, as others have stated, rodents should not be blamed for the recent failure of developing new sepsis therapies [14]. Sepsis is a highly heterogeneous condition and involves multiple immune pathways that can be both helpful and harmful when activated at different levels and during different stages of the disease. A precision approach may be needed to develop effective sepsis therapies. One should also remember that animal research was instrumental in developing antibiotics, which play a central role in sepsis treatment. In 1945, the Nobel Prize was awarded to Ernst Chain and Howard Florey "for the discovery of penicillin and its curative effect in various infectious diseases." Their seminal experiments showed that penicillin could cure mice infected with lethal doses of bacteria [15]. Mice remain a useful model to investigate the cellular and molecular processes that contribute to sepsis pathogenesis. However, we should constantly strive to refine our animal models of sepsis to better mimic the human condition, as well as work toward developing complementary non-animal models. When implementing the models described in this book, investigators are advised to follow the recent recommendations Minimum Quality Thresholds in Preclinical Sepsis Studies (MQTiPSS) [16].

This volume of the *Methods in Molecular Biology* series presents a variety of animal models that are commonly used to study sepsis and some key procedures to measure specific disease outcomes. The chapters describe well-established surgical and nonsurgical rodent models of sepsis, presented by experts in the field. In addition, the book includes protocols for burn injury and sepsis, modeling the mouse intensive care unit (MICU), and the development of humanized mice, which may be useful tools to increase the translational potential of rodent sepsis research. There is a chapter discussing the use of biomarkers for sepsis diagnosis and prognosis (in humans and mice), as well as two chapters describing specific methods for biomarker measurement. The book would not be complete without a chapter describing the use of analgesics and humane endpoints in rodent sepsis research. Finally, agent-based computational modeling is presented as a valuable complementary approach to study sepsis.

I would like to express my gratitude to the people who have helped make this book successful. I thank the authors who have worked hard to develop the individual chapters and have been very patient and responsive during the editing phase. I thank the series editor John Walker for his encouragement and guidance. I thank my institution, Texas Tech University Health Sciences Center at El Paso, for providing an excellent environment for my research and scholarship. Finally, I am greatly indebted to my husband and family for their constant support.

El Paso, TX, USA ***Wendy E. Walker***

References

1. Singer M, Deutschman CS, Seymour CW et al (2016) The third international consensus definitions for sepsis and septic shock (Sepsis-3). JAMA 315:801–810
2. Delano MJ, Ward PA (2016) The immune system's role in sepsis progression, resolution, and long-term outcome. Immunol Rev 274:330–353
3. Ayala A, Chaudry IH (1996) Immune dysfunction in murine polymicrobial sepsis: mediators, macrophages, lymphocytes and apoptosis. Shock 6 Suppl 1:S27–S38
4. Gentile LF, Cuenca AG, Efron PA et al (2012) Persistent inflammation and immunosuppression: a common syndrome and new horizon for surgical intensive care. J Trauma Acute Care Surg 72:1491–1501
5. Ward NS, Casserly B, Ayala A (2008) The compensatory anti-inflammatory response syndrome (CARS) in critically ill patients. Clin Chest Med 29:617–625, viii
6. Liu V, Escobar GJ, Greene JD et al (2014) Hospital deaths in patients with sepsis from 2 independent cohorts. JAMA 312:90–92
7. Huang CY, Daniels R, Lembo A et al (2019) Life after sepsis: an international survey of survivors to understand the post-sepsis syndrome. Int J Qual Health Care 31:191–198
8. Nasa P, Juneja D, Singh O (2012) Severe sepsis and septic shock in the elderly: an overview. World J Crit Care Med 1:23–30
9. Sakr Y, Jaschinski U, Wittebole X et al (2018) Sepsis in intensive care unit patients: worldwide data from the intensive care over nations audit. Open Forum Infect Dis 5:ofy313
10. Dellinger RP, Levy MM, Carlet JM et al (2008) Surviving sepsis campaign: international guidelines for management of severe sepsis and septic shock: 2008. Crit Care Med 36:296–327
11. Hatfield KM, Dantes RB, Baggs J et al (2018) Assessing variability in hospital-level mortality among US medicare beneficiaries with hospitalizations for severe sepsis and septic shock. Crit Care Med 46:1753
12. Seok J, Warren HS, Cuenca AG et al (2013) Genomic responses in mouse models poorly mimic human inflammatory diseases. Proc Natl Acad Sci U S A 110:3507–3512
13. Takao K, Miyakawa T (2015) Genomic responses in mouse models greatly mimic human inflammatory diseases. Proc Natl Acad Sci U S A 112:1167–1172
14. De Maio A (2020) Do not blame the rodent for the failure of developing sepsis therapies. Shock 54:631–632
15. Chain E, Florey HW, Gardner AD et al (1940) Penicillin as a chemotherapeutic agent. Lancet 236:226–228
16. Zingarelli B, Coopersmith CM, Drechsler S et al (2019) Part I: minimum quality threshold in preclinical sepsis studies (MQTiPSS) for study design and humane modeling endpoints. Shock 51:10–22

Contents

Contributors

GARY AN • *Department of Surgery, University of Vermont, Burlington, VT, USA*
CHRISTINE A. BOEHM • *Laboratory Animal Research Center, Texas Tech University Health Sciences Center at El Paso, El Paso, TX, USA*
JULIA K. BOHANNON • *Department of Anesthesiology, Vanderbilt University Medical Center, Nashville, TN, USA; Department of Pathology, Microbiology, and Immunology, Vanderbilt University Medical Center, Nashville, TN, USA*
R. CHASE COCKRELL • *Department of Surgery, University of Vermont, Burlington, VT, USA*
CRAIG M. COOPERSMITH • *Department of Surgery and Emory Critical Care Center, Emory University School of Medicine, Atlanta, GA, USA*
ANA CUENDA • *Department of Immunology and Oncology, Centro Nacional de Biotecnología/CSIC, Madrid, Spain*
MEIHONG DENG • *Department of Surgery, The Ohio State University, Columbus, OH, USA*
SUSANNE DRECHSLER • *Ludwig Boltzmann Institute for Experimental and Clinical Traumatology in the AUVA Research Center, Vienna, Austria*
PHILIP A. EFRON • *Department of Surgery, University of Florida College of Medicine, Gainesville, FL, USA*
PILAR FAJARDO • *Department of Immunology and Oncology, Centro Nacional de Biotecnología/CSIC, Madrid, Spain*
SAILAJA GHANTA • *Department of Pediatric Newborn Medicine, Brigham and Women's Hospital and Harvard Medical School, Boston, MA, USA*
QINGQING GONG • *Division of Newborn Medicine, Department of Pediatrics, Washington University School of Medicine, St. Louis Children's Hospital, St. Louis, MO, USA*
MISTY GOOD • *Division of Newborn Medicine, Department of Pediatrics, Washington University School of Medicine, Louis Children's Hospital, St. Louis, MO, USA*
DINESH G. GOSWAMI • *Center of Emphasis in Infectious Diseases, Department of Molecular and Translational Medicine, Paul L Foster School of Medicine, Texas Tech University Health Sciences Center at El Paso, El Paso, TX, USA*
MICHAEL GRÖGER • *Institute for Anesthesiological Pathophysiology and Process Engineering, Ulm University Medical Center, Ulm, Germany*
XIAN-HUI HE • *Department of Immunobiology, College of Life Science and Technology, Jinan University, Guangzhou, China*
ERICA L. HEIPERTZ • *Center of Emphasis in Infectious Diseases, Department of Molecular and Translational Medicine, Paul L Foster School of Medicine, Texas Tech University Health Sciences Center at El Paso, El Paso, TX, USA*
ANNA HERMINGHAUS • *Department of Anesthesiology, University Hospital Duesseldorf, Duesseldorf, Germany*
ANTONIO HERNANDEZ • *Department of Anesthesiology, Vanderbilt University Medical Center, Nashville, TN, USA*
YACHANA KATARIA • *Department of Pathology and Laboratory Medicine, Boston School of Medicine, Boston, MA, USA*
SANDRA KRESS • *Institute for Anesthesiological Pathophysiology and Process Engineering, Ulm University Medical Center, Ulm, Germany*

MIN-YOUNG KWON • *Department of Pediatric Newborn Medicine, Brigham and Women's Hospital and Harvard Medical School, Boston, MA, USA*
WYATT E. LANIK • *Division of Newborn Medicine, Department of Pediatrics, Washington University School of Medicine, St. Louis Children's Hospital, St. Louis, MO, USA*
SHAWN D. LARSON • *Department of Surgery, University of Florida College of Medicine, Gainesville, FL, USA*
BRIAN W. LEBLANC • *Division of Surgical Research, Department of Surgery, Rhode Island Hospital, Providence, RI, USA; Warren Alpert Medical School, Brown University, Providence, RI, USA*
CRAIG T. LEFORT • *Division of Surgical Research, Department of Surgery, Rhode Island Hospital, Providence, RI, USA; Warren Alpert Medical School, Brown University, Providence, RI, USA*
OSCAR MCCOOK • *Institute for Anesthesiological Pathophysiology and Process Engineering, Ulm University Medical Center, Ulm, Germany*
TINA S. MELE • *Department of Surgery, Western University, London, ON, Canada*
TAMARA MERZ • *Institute for Anesthesiological Pathophysiology and Process Engineering, Ulm University Medical Center, Ulm, Germany*
BELGACEM MIHI • *Division of Newborn Medicine, Department of Pediatrics, Washington University School of Medicine, St. Louis Children's Hospital, St. Louis, MO, USA*
LYLE L. MOLDAWER • *Department of Surgery, University of Florida College of Medicine, Gainesville, FL, USA*
JEAN A. NEMZEK • *Unit for Laboratory Animal Medicine, University of Michigan, Ann Arbor, MI, USA*
TAKEHIKO OAMI • *Department of Surgery and Emory Critical Care Center, Emory University School of Medicine, Atlanta, GA, USA; Department of Emergency and Critical Care Medicine, Chiba University Graduate School of Medicine, Chiba, Japan*
MARCIN OSUCHOWSKI • *Ludwig Boltzmann Institute for Experimental and Clinical Traumatology in the AUVA Research Center, Vienna, Austria*
DONG-YUN OUYANG • *Department of Immunobiology, College of Life Science and Technology, Jinan University, Guangzhou, China*
DIMITRI PAPPAS • *Department of Chemistry and Biochemistry, Texas Tech University, Lubbock, TX, USA*
NAEEM K. PATIL • *Department of Anesthesiology, Vanderbilt University Medical Center, Nashville, TN, USA*
MARK A. PERRELLA • *Department of Pediatric Newborn Medicine, Brigham and Women's Hospital and Harvard Medical School, Boston, MA, USA; Division of Pulmonary and Critical Care Medicine, Department of Medicine, Brigham and Women's Hospital and Harvard Medical School, Boston, MA, USA*
OLAF PICKER • *Department of Anesthesiology, University Hospital Duesseldorf, Duesseldorf, Germany*
PETER RADERMACHER • *Institute for Anesthesiological Pathophysiology and Process Engineering, Ulm University Medical Center, Ulm, Germany*
MIREYA G. RAMOS MUNIZ • *Department of Biological Sciences, University of Texas at El Paso, El Paso, TX, USA*
DANIEL REMICK • *Department of Pathology and Laboratory Medicine, Boston School of Medicine, Boston, MA, USA*
JAIMAR C. RINCON • *Department of Surgery, University of Florida College of Medicine, Gainesville, FL, USA*

MICHELLE A. SANCHEZ GUILLEN • *Department of Biological Sciences, University of Texas at El Paso, El Paso, TX, USA*
JUAN JOSÉ SANZ-EZQUERRO • *Department of Molecular and Cellular Biology, Centro Nacional de Biotecnología/CSIC, Madrid, Spain*
NICOLE R. SETZU • *Department of Biological Sciences, University of Texas at El Paso, El Paso, TX, USA*
CHARLES T. SPENCER • *Department of Biological Sciences, University of Texas at El Paso, El Paso, TX, USA*
WENDY E. WALKER • *Department of Molecular and Translational Medicine, Texas Tech University Health Sciences, Center at El Paso, El Paso, TX, USA*
HUI XU • *State Key Laboratory of Oral Diseases, National Clinical Research Center for Oral Diseases, Department of Orthodontics, West China Hospital of Stomatology, Sichuan University, Chengdu, P. R., China*
LI-HUI XU • *Department of Cell Biolog, College of Life Science and Technology, Jinan University, Guangzhou, China*
YIJIA YANG • *Department of Chemistry and Biochemistry, Texas Tech University, Lubbock, TX, USA*
YUN ZHOU • *Department of Chemistry and Biochemistry, Texas Tech University, Lubbock, TX, USA*

Chapter 1

Cecal Ligation and Puncture

Susanne Drechsler and Marcin Osuchowski

Abstract

Cecal ligation and puncture (CLP) is referred to as the "gold standard" rodent model for abdominal sepsis. CLP creates a continuously leaking, polymicrobial infectious focus in the abdomen. The abdominal cavity is opened under general anesthesia and analgesia and the cecum is exposed, ligated underneath the ileocecal valve, and punctured with a needle. A small amount of feces is pressed out through the puncture and the cecum is repositioned into the abdomen, which is then closed with single button sutures and tissue glue. CLP severity can be influenced via the length of the ligated cecum as well as the needle size. Within 24 h, animals develop clinical signs of a systemic bacterial infection. Analgesia, wide range antibiotic treatment, and fluid resuscitation should be administered during the acute phase of sepsis to increase the clinical relevance of the CLP model.

Key words Sepsis, Small animal model, CLP, Polymicrobial, Peritonitis

1 Introduction

Every year, approximately 31.5 million patients suffer from sepsis and 5.3 million of them do not survive [1]. Due to the variability in the anatomical origin of sepsis (e.g., most frequently the lungs, abdomen, or urinary tract) and the heterogenic dynamics of its progression, timely diagnosis and treatment of this disease are ongoing challenges for clinicians and scientists. Despite recent decades of intensive research and a better understanding of the underlying pathophysiology, the standard treatment of septic patients remains nonspecific and expensive and is mainly comprised of broad spectrum antibiotics and fluid resuscitation [2]. Early goal directed therapy (EGDT) was not shown to be effective in recent trials [3, 4]. Abdominal sepsis is one of the most common forms of sepsis [5], in which the bowel acts as a source for polymicrobial infection [6]. Abdominal sepsis typically occurs when gastrointestinal content leaks into the abdominal cavity as a second-

Wendy E. Walker (ed.), *Sepsis: Methods and Protocols*, Methods in Molecular Biology, vol. 2321,
https://doi.org/10.1007/978-1-0716-1488-4_1, © Springer Science+Business Media, LLC, part of Springer Nature 2021

ary complication after abdominal surgery and traumatic injury, but is also associated with the spontaneous rupture of gastric ulcers [7]. The host typically responds with an initially localized inflammatory/immune reaction and the development of peritonitis. If fecal microorganisms and/or their components then enter the circulation, they trigger a strong systemic inflammatory response which can become dysregulated and may result in a life-threatening condition of impaired organ function/injury referred to as sepsis [8].

Rodent models that simulate these immune-inflammatory processes are a major pillar of sepsis research. Today, CLP is considered to be the "gold standard" model for recapitulating the main features of human immune reactions in polymicrobial abdominal sepsis [9]. It consists of a laparotomy, followed by the ligation of the cecum underneath the ileocecal valve and the puncture of the cecum with a needle. This creates a focal necrosis and continuous leakage of fecal material. The method was developed almost 40 years ago by Wichterman et al. [10], who modified a previously published model of cecal ligation in rats [11]. After CLP, clinical signs of sepsis occur within 12–24 h depending on its severity. Septic rodents show typical symptoms of sepsis such as piloerection, tachypnea, tachycardia, reduced vigilance and mobility, decreased food and water intake, weight gain/loss (depending on the phase of sepsis), and a change in body temperature. In mice, CLP induced sepsis is typically associated with hypothermia, while rats maintain their core temperature [12].

The severity of the developing disease ranges from mild (chronic) to severe (acute) sepsis and can be influenced via the length of the ligated cecum [13], as well as the needle size (typically ranging from 26 to 18 gauge) and the number of punctures (typically two). Given that the severity grade of the CLP procedure modifies both the underlying pathophysiology and the disease outcome, high consistency in performing this model is a fundamental prerequisite for successful application of the CLP protocol in a lab [14]. However, the susceptibility to CLP-induced sepsis varies between rodent strains and is dependent on age and gender, resulting in significant inter-lab variability. Preliminary optimization studies are helpful to establish the protocol (*see* **Note 1**). Today, CLP is mostly used in mice and rats, but it has also been adapted for use in pigs [15] and dogs [16].

While the original protocol lacked any standardized supportive medication, in the recent decades, the CLP method has been refined with regard to animal welfare and its similarity to the clinical situation by the implementation of appropriate analgesics, broad spectrum antibiotic treatment, and fluid resuscitation. Opioids, β-lactam antibiotics (from the carbapenem family), and body-temperature iso-osmolar fluids are the treatments of choice. Even so, CLP induced sepsis cannot fully recapitulate all pathophysiological and immunological features of human sepsis. One key reason is that CLP is typically performed in genetically homogeneous, healthy, young rodents of a single gender, which cannot properly represent heterogenic human patient cohorts with their wide variations in genetic background, age, endocrine and immune status, pre-existing comorbidities, and ongoing treatments. Recently published expert consensus recommendations on Minimum Quality Thresholds in Preclinical Sepsis Studies (MQTiPSS) [17–19] bring awareness to the do's and don'ts that should be observed in sepsis experimentation (including the CLP model) to maximize their translatability.

2 Materials

2.1 CLP

1. Medical drape.
2. 1 mL syringes.
3. 26G needles for s.c. injection.
4. needles for cecum puncture (26–18 gauge range).
5. Buprenorphine injection solution (*see* **Note 2**).
6. Ringer's lactate solution.
7. Warming devices for intraoperative and postoperative thermal support (see **Note 3**).
8. Glass bead dry sterilizer (optional, *see* **Note 4**).
9. Inhalation anesthesia equipment (*see* **Note 5**).
10. Hair clipper.
11. 10% povidone–iodine solution.
12. Sterile gauze swabs, 5 × 5 cm.
13. Spray flask with 70% ethanol.
14. Surgical cup filled with H_2O.
15. Nonsterile and sterile gloves.
16. Eye ointment.
17. Surgical suture material, silk, USP 3/0 with circular needle.
18. Surgical suture material, silk, USP 3/0 on a reel.
19. Small surgical scissors.
20. Small anatomical forceps.
21. Small surgical forceps.
22. Needle holder.
23. Tissue glue.

3 Methods

3.1 Preparation for Surgery

1. Put on nonsterile gloves.
2. Prepare the warming device (*see* **Note 3**).
3. Cover the surgical table with sterile medical drape.
4. Preheat the glass bead dry sterilizer (*see* **Note 4**).
5. Fill a cup with H_2O (to clean the surgical instruments).
6. Sterilize the surgical instruments using an autoclave or glass bead dry sterilizer (*see* **Note 4**).
7. Prepare suture material, needles, and tissue glue.

3.2 CLP Surgery

1. Measure the mouse body weight.
2. Inject the mouse with buprenorphine (0.05–0.01 mg/kg) in Ringer's lactate solution (0.025 ml/g BW) subcutaneously, using a 26G needle (*see* **Note 2**) and wait for 30 min.
3. Anesthetize the animal (*see* **Note 5**).
4. Apply eye ointment.
5. Shave the mouse's abdomen.
6. Put the animal on the warming device in a supine position.
7. Disinfect the shaved skin with povidone–iodine solution, followed by ethanol on a gauze swab.
8. Switch to sterile gloves.
9. Make a 1 cm skin incision in the midline of the abdomen using the surgical forceps and scissors (Fig. 1a).
10. Identify the linea alba and lift it with the surgical forceps.
11. Carefully cut a 1 cm incision in the linea alba to open the abdomen, using the scissors (*see* **Note 6**).
12. Switch from scissors to anatomical forceps and identify the cecum on the left side of the abdomen (from the animal's perspective) (*see* **Note 7**).
13. Carefully pull the cecum out of the abdomen, and position it so you can see the ileocecal valve (Fig. 1b, *see* **Note 8**).
14. Cut off a 10 cm silk thread from the spool and ligate the cecum underneath the ileocecal valve with two single button knots (Fig. 1c, *see* **Note 9**).
15. Puncture the cecum at the basis and the apex ceci with the chosen needle (Fig. 1d, *see* **Note 10**).
16. Very carefully press out a little bit of fecal content (1 mm) through the holes with your fingers or with the anatomical forceps to ensure patency of the puncture.

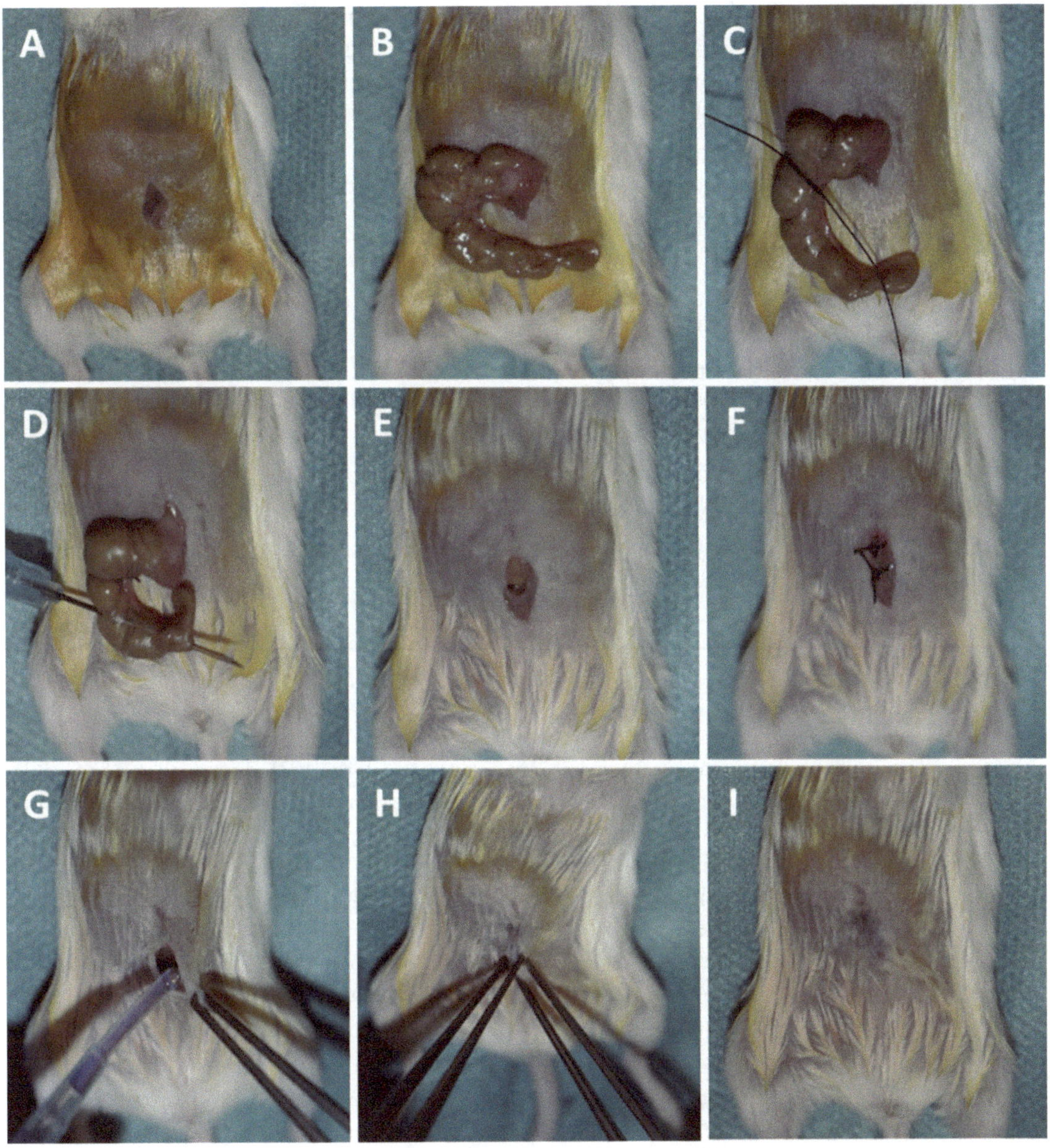

Fig. 1 CLP Procedure: (**a**) skin incision, (**b**) exteriorization of the cecum, (**c**) ligation of the cecum, (**d**) puncture of the cecum, (**e**) repositioning of the cecum, (**f**) abdominal wall closure with single button sutures, (**g + h**) skin closure with tissue glue, (**i**) closed skin, postoperatively

17. Use the anatomical forceps to reposition the cecum back into the abdominal cavity but pay attention that no fecal content adheres to the wound margin (Fig. 1e).
18. Close the abdomen with single simple interrupted sutures using a needle holder and suture material (Fig. 1f, *see* **Note 11**).

19. Carefully apply a small drop of tissue glue to the inside proximal surfaces of the skin (Fig. 1g) and rapidly adjust and close the skin cut using the anatomical forceps (Fig. 1h + i, see **Note 12**).
20. Put the animal in a separate cage and keep its body temperature between 36.5 and 37.5 °C, using thermal support, until it is fully awake (*see* **Note 3**).
21. Clean the surgical instruments with H_2O.
22. Sterilize surgical instruments in the glass bead sterilizer (*see* **Note 4**).
23. Spray your hands with ethanol (*see* **Note 13**).
24. Continue with the next mouse.

3.3 Postoperative Care

1. Inject broad spectrum antibiotics at 2 h after CLP (*see* **Note 14**).
2. Administer additional doses of analgesics for pain relief (*see* **Note 15**).

4 Notes

1. Establishing the CLP model of a specified severity grade for survival studies can require a relatively large number of animals. For a meaningful result, we recommend starting with 10 mice and observing their outcome for at least 7 days post-CLP. The protocol (ligated portion and puncture size) can then be slightly adapted before repeating the experiment with the next animal group. However, once the severity level is established, it is recommended to extend survival studies to 14–28 days. An experienced surgeon should be able to perform CLP within 7–10 min.
2. For continuous analgesia, buprenorphine should be administered every 6-8 h for at least the first 3-5 days starting at least 30 min before the first CLP surgery. Be aware that buprenorphine has depressive respiratory effects.
3. Devices which cannot overheat the animal (e.g., circulating water blankets) are suitable for intraoperative thermal support. A heat lamp is useful for postoperative thermal support, but requires frequent monitoring of body temperature to avoid overheating.
4. A benchtop sterilizer for surgical instruments is useful if more than one animal will undergo CLP. Otherwise, instruments can be sterilized in an autoclave. Putting contaminated surgical instruments into ethanol should be avoided [20].

5. Alternatively, intraperitoneal injection anesthesia can be used. However, inhalation anesthesia is best suited for very short surgeries like CLP, because this method ensures a rapid recovery time. Your institutional IACUC and local regulatory authorities can provide guidance on appropriate anesthetic drugs and their respective doses.
6. If the peritoneal membrane is not cut along the *linea alba*, bleeding will likely occur.
7. Although uncommon, in some mice the cecum is positioned deep inside the abdominal cavity, or on the animal's right side. If so, manipulation of the gut should never be performed with surgical forceps, instead anatomical forceps should be used. When looking for the cecum, take care not to injure the gut and/or small mesenteric blood vessels.
8. The cecum may become twisted, and should be carefully unfolded prior to its ligation.
9. To change the severity of CLP, a smaller portion of cecum can be ligated (*see* **Note 10**). However, the ligated portion must always remain the same in a given study (e.g., two-thirds of the cecum).
10. The number of punctures depends on the desired sepsis severity. It is possible, for example, to puncture the cecum at two different positions or make a puncture through both walls of the cecum (referred to as a through-and-through puncture).
11. Alternatively, running sutures can be used.
12. Tissue glue is optional; the skin can also be closed with sutures or wound clips. When closing the skin with glue, it is prudent to use a very small amount of tissue glue and apply it only to the inside surface of the skin; otherwise the forceps may become glued to the skin.
13. Alternatively, any other sterilizing fluid or a fresh pair of gloves can be used.
14. A standardized regimen for broad spectrum antibiotic treatment (e.g., imipenem, meropenem) should be implemented after CLP. Various antibiotic treatment protocols can be applied depending on the study design.
15. Depending on the strain and/or sex, analgesia and adequate fluid resuscitation should be continued for the first 3–5 days after CLP [21] and whenever an animal deteriorates at a later time. Implementation/use of scoring for humane endpoint identification is recommended, both existing examples [22–24] and/or your own scoring protocols can be applied.

References

1. Fleischmann C, Scherag A, Adhikari NK et al (2016) Assessment of global incidence and mortality of hospital-treated sepsis. Current estimates and limitations. Am J Respir Crit Care Med 193:259–272
2. Coopersmith CM, De Backer D, Deutschman CS et al (2018) Surviving sepsis campaign: research priorities for sepsis and septic shock. Intensive Care Med 44:1400–1426
3. Chen X, Zhu W, Tan J et al (2017) Early outcome of early-goal directed therapy for patients with sepsis or septic shock: a systematic review and meta-analysis of randomized controlled trials. Oncotarget 8:27510
4. Zhang L, Zhu G, Han L et al (2015) Early goal-directed therapy in the management of severe sepsis or septic shock in adults: a meta-analysis of randomized controlled trials. BMC Med 13:71
5. Mayr FB, Yende S, Angus DC (2014) Epidemiology of severe sepsis. Virulence 5:4–11
6. Angus DC, Van der Poll T (2013) Severe sepsis and septic shock. N Engl J Med 369:840–851
7. Alverdy J, Hyoju S, Weigerinck M et al (2017) The gut microbiome and the mechanism of surgical infection. Br J Surg 104:e14–e23
8. Taeb AM, Hooper MH, Marik PE (2017) Sepsis: current definition, pathophysiology, diagnosis, and management. Nutr Clin Pract 32:296–308
9. Dejager L, Pinheiro I, Dejonckheere E et al (2011) Cecal ligation and puncture: the gold standard model for polymicrobial sepsis? Trends Microbiol 19:198–208
10. Wichterman KA, Baue AE, Chaudry IH (1980) Sepsis and septic shock—a review of laboratory models and a proposal. J Surg Res 29:189–201
11. Ryan NT, Blackburn GL, Clowes GH Jr (1974) Differential tissue sensitivity to elevated endogenous insulin levels during experimental peritonitis in rats. Metabolism 23:1081–1089
12. Zolfaghari PS, Pinto BB, Dyson A et al (2013) The metabolic phenotype of rodent sepsis: cause for concern? Intensive Care Med Exp 1:6
13. Singleton K, Wischmeyer P (2003) Distance of cecum ligated influences mortality, tumor necrosis factor-alpha and interleukin-6 expression following cecal ligation and puncture in the rat. Eur Surg Res 35:486–491
14. Rittirsch D, Huber-Lang MS, Flierl MA et al (2009) Immunodesign of experimental sepsis by cecal ligation and puncture. Nat Protoc 4:31–36
15. Kieslichova E, Rocen M, Merta D et al (2013) The effect of immunosuppression on manifestations of sepsis in an animal model of cecal ligation and puncture. Transplant Proc 45 (2):770–777
16. Stahl TJ, Alden PB, Ring WS et al (1990) Sepsis-induced diastolic dysfunction in chronic canine peritonitis. Am J Phys Heart Circ Phys 258:H625–H633
17. Zingarelli B, Coopersmith CM, Drechsler S et al (2019) Part I: minimum quality threshold in preclinical sepsis studies (MQTiPSS) for study design and humane modeling endpoints. Shock 51:10–22
18. Libert C, Ayala A, Bauer M et al (2019) Part II: minimum quality threshold in pre-clinical sepsis studies (MQTiPSS) for types of infections and organ dysfunction endpoints. Shock 51:23
19. Hellman J, Bahrami S, Boros M et al (2019) Part III: minimum quality threshold in preclinical sepsis studies (MQTiPSS) for fluid resuscitation and antimicrobial therapy endpoints. Shock 51:33–43
20. de Melo CD, de Oliveira Lopes LK, Hu H et al (2017) Alcohol fixation of bacteria to surgical instruments increases cleaning difficulty and may contribute to sterilization inefficacy. Am J Infect Control 45:e81–e86
21. Iskander KN, Vaickus M, Duffy ER et al (2016) Shorter duration of post-operative antibiotics for Cecal ligation and puncture does not increase inflammation or mortality. PLoS One 11:e0163005
22. Huet O, Ramsey D, Miljavec S et al (2013) Ensuring animal welfare while meeting scientific aims using a murine pneumonia model of septic shock. Shock 39:488–494
23. Shrum B, Anantha RV, Xu SX et al (2014) A robust scoring system to evaluate sepsis severity in an animal model. BMC Res Notes 7:233
24. Nemzek JA, Hugunin KM, Opp MR (2008) Modeling sepsis in the laboratory: merging sound science with animal well-being. Comp Med 58:120–128

Chapter 2

Colon Ascendens Stent Peritonitis (CASP)

Anna Herminghaus and Olaf Picker

Abstract

Colon ascendens stent peritonitis (CASP) is one of the well-established models of experimental abdominal sepsis. In CASP surgery, an open link between the gut lumen and the abdominal cavity is created by placing a stent into the colon ascendens. This mimics well the insufficient intestinal anastomosis. It causes a continuous leakage of the gut contents into the peritoneum and leads therefore to peritonitis and sepsis. The abdominal cavity is opened under general anesthesia and a plastic stent is located through and sutured to the colonic wall. The septic severity in CASP models can be titrated by altering the size of the stent catheter. Therefore, CASP models with small stents sizes are suitable for long-term studies and studies with mild/moderate sepsis severity. Within 24 h, animals develop clinical signs of sepsis. Monitoring of the clinical state, sufficient analgesia, appropriate antibiotics and fluid resuscitation should be performed postoperatively.

Key words Sepsis, CASP, Abdominal infection, Peritonitis

1 Introduction

Complicated intraabdominal infections are a relevant clinical problem, and when poorly managed, they cause a high mortality (9.2%) [1]. The mortality in postoperative peritonitis is even higher (22–55%) [2]. Abdominal sepsis has a special clinical relevance, even if it is not a leading focus of sepsis [3]. The abdomen may be implicated as the primary source of infection during cholecystitis, diverticulitis or bowel ischemia, and as a secondary source, for example due to anastomotic leak or postoperative abscess, or intestinal ischemia resulting from splanchnic hypoperfusion [4].

Apart from its role as an initiator of sepsis, the gut can also be a victim of dysregulated inflammatory processes during sepsis. It is well known that sepsis results in microcirculatory dysfunction, increased intestinal permeability, and as a consequence, bacterial translocation through the mucosa [5]. Our current knowledge about the pathophysiology of sepsis has been largely obtained from rodent models [6].

Wendy E. Walker (ed.), *Sepsis: Methods and Protocols*, Methods in Molecular Biology, vol. 2321,
https://doi.org/10.1007/978-1-0716-1488-4_2,

There are many experimental models of sepsis but none of them perfectly reflect the real clinical scenario. In colon ascendens stent peritonitis (CASP) surgery, an open link is created between the gut lumen and the abdominal cavity by placing a stent into the colon ascendens.

This provides an excellent mimic of insufficient intestinal anastomosis. It causes a continuous leakage of the gut contents into the peritoneal cavity and leads therefore to peritonitis and sepsis. CASP is a well-established model to study sepsis experimentally in mice. Lustig et al. adjusted and transformed the model to rats [7]. Mice are a valuable tool for experimental studies considering the knockout technology available. However, over the past few decades, the physiology of all major organ systems has been investigated to a greater degree in rats versus mice. In addition, studies on organ function are easier to perform and less prone to errors in rats than in mice for anatomical reasons [7]. Furthermore, the severity of the septic insult in CASP models can be titrated by altering the size of the stent catheter [7]. Therefore, CASP models with a small stent size are suitable for long term studies and studies which focus on reversible abdominal sepsis or mild to moderate sepsis severity [8]. The CASP model appears to reproduce the immunological response as well as most organ dysfunctions observed in human sepsis [8, 9]. In CASP, as in every other experimental sepsis model, recently published expert consensus recommendations on minimum quality thresholds in preclinical sepsis studies (MQTiPSS) [10] should be considered. Nevertheless, the transfer of the results obtained in rodents to human patient trials must be performed very carefully.

2 Materials

1. 1 ml and 10 ml syringes.
2. Nonsterile and sterile gloves.
3. Needles for subcutaneous injection (27 G).
4. Peripheral venous catheters (size depending on the estimated severity of sepsis 14–18 G).
5. Gauze compresses: 10 × 10 cm cotton swabs.
6. Warming device (*see* **Note 1**).
7. Inhalation anesthesia equipment (*see* **Note 2**).
8. Buprenorphine injection solution (*see* **Note 3**).
9. Ringer Lactate Solution (*see* **Note 4**).
10. Skin disinfection solution in a spray flask.

11. Eye ointment.
12. Surgical suture material: Prolene, USP 6/0 with circular needle.
13. Surgical suture material: vicryl, antibacterial, USP 4/0 with circular needle.
14. Small surgical scissors.
15. Small surgical forceps.
16. Needle holder.
17. Sterile surgical fenestrated drape.

3 Methods

Standard protocol for CASP in rats

1. Prepare everything you need for the operation: suture material, chirurgical sterile drapes, peripheral venous catheter(s), cotton swabs, prewarmed Ringer solution in a 10 ml syringe, surgical instruments (*see* **Note 5**).
2. Put on nonsterile gloves.
3. Weigh the animal.
4. Inject buprenorphine subcutaneously (with a needle 27 G) (dose: 0.05–0.01 mg/kg) in Ringer Lactate Solution (0.0025 ml/g BW) (*see* **Note 3**).
5. Prepare the warming device (*see* **Note 1**).
6. Anaesthetize the animal with a volatile anesthetic (e.g., sevoflurane) (*see* **Note 2**).
7. Apply eye ointment.
8. Put the animal on the operating table in a supine position, maintaining anesthesia.
9. Prepare the stent: using the scissors, make some small superficial cuts on the plastic surface of the catheter 8–10 mm from the distal end to enable it to be fixed in place later with the 6.0 suture material. Wet the cotton swabs with the Ringer Lactate Solution.
10. Apply the disinfection solution to the skin (*see* **Note 6**).
11. Switch to sterile gloves.
12. Cover the rat with the sterile fenestrated surgical drape (arranging the hole over the center of the abdomen).
13. Make a 2 cm skin incision in the midline of the abdomen using the surgical forceps and scissors, identify the linea alba and open the muscle sheet along the linea alba.
14. Identify the colon using cotton swabs (*see* **Note 7**).

15. Carefully pull the colon out of the abdomen, identify the ileocecal valve and the proximal colon (*see* **Note 8**).
16. Place a suture (6.0) in the colon wall directly beyond the ileocecal valve. This will allow the catheter to be attached later on. Prepare a loop in the suture material to secure the catheter.
17. Puncture the colon through the loop with the peripheral venous catheter. As soon as you penetrate the colon lumen with the needle, withdraw the needle a little bit und push the plastic catheter about 8 mm into the colon lumen.
18. Secure the catheter with the 6.0 suture (*see* **Note 9**).
19. To verify a successful penetration, press out a small amount of the fecal contents through the catheter, using your fingers.
20. Replace the colon back into the abdominal cavity using the cotton swabs.
21. Apply prewarmed Ringer solution (0.015 ml/g BW) into the abdominal cavity for fluid resuscitation (*see* **Note 4**).
22. Close the peritoneal membrane and associated muscle sheet with a running suture (4.0). Close the skin with single simple interrupted sutures (4.0) (*see* **Note 10**).
23. Remove any blood from the skin and dry it as much as possible.
24. Place the animal in a separate cage and monitor until it is fully conscious (*see* **Note 11**).
25. Clean the surgical instruments with H_2O and sterilize them.
26. Monitor the operated animal every 4–6 h to prevent exorbitant suffering (see **Note 12**). The Modified Septic Rat Severity Score (SRSS) can be used for this purpose (Table 1).
27. Administer additional fluid resuscitation (*see* **Note 4**) and antibiotics (*see* **Note 13**).

Table 1
Modified Septic Rat Severity score (SRSS) [12, 13]

Examination	Results	Points
Body weight	1. Initial weight (iw) ___g 2. End weight (ew) ___g 3. Weight loss___%	% < 5 ⇒ 0 P % < 15 ⇒ 2 P % < 20 ⇒ 3 P % > 20 ⇒ 10 P
Appearance	1. Normal appearance, fur smooth, clean 2. Slight grooming deficiency, rough fur 3. Increasing grooming deficiency, rings around eyes, anus 4. Clear grooming deficiency, crusty eyes, bedding sticks to anus	⇒ 0 P ⇒ 1 P ⇒ 2 P ⇒ 3 P

(continued)

Table 1
(continued)

Examination	Results	Points
Spontaneous behavior	1. Rat investigates cage, active	⇒ 0 P
	2. Rat remains in one place, movement of entire body present	⇒ 1 P
	3. Hunched posture, swaying gait	⇒ 3 P
	4. Immobile, lateral position	⇒ 10 P
Provoked behavior	1. Rat flees when opening cage, strong muscle tonus	⇒ 0 P
	2. Rat flees when hand approaches	⇒ 1 P
	3. Rat flees when touched	⇒ 3 P
	4. No flight reaction	⇒ 10 P
Expiratory breathing sound	No	⇒ 0 P
	Yes	⇒ 1 P
Abdominal palpation	1. No pain when applying pressure, soft abdomen	⇒ 0 P
	2. Slight reaction to abdominal palpation, soft abdomen	⇒ 1 P
	3. Clear pain reaction to abdominal palpation, abdominal resistance	⇒ 3 P
	4. Clear pain reaction to abdominal palpation, hard abdomen	⇒ 10 P
Condition of droppings	1. A lot of normal droppings in cage, defecating during examination	⇒ 0 P
	2. A lot of droppings in cage, droppings with blood, runny or mucous	⇒ 1 P
	3. Few droppings in cage, independent of the condition	⇒ 2 P
	4. No droppings in cage	⇒ 10 P

4 Notes

1. A heating lamp or a heating plate are suitable for the operation. Another simple makeshift heating device is normal gloves filled with warm water, positioned on both sides of the animal. Cooling is not a concern, because the CASP surgery can performed by a well-trained investigator within 10 min.
2. A sevoflurane concentration of 3.0 Vol%, FiO_2 0.5 is recommended to maintain spontaneous respiration. As alternative, you can use pentobarbital (intraperitoneal, 60 mg/kg) injection, but you have to take into consideration that the emergence from anesthesia after a single dose of pentobarbital takes much longer compared to inhalation anesthesia with sevoflurane (3–4 h vs. 5 min respectively) resulting in a longer monitoring of the animal.

3. For an effective postoperative analgesia, it is recommended to administer Buprenorphine every 6–8 h for at least 5 days.
4. Iso-osmolar balanced crystalloid solutions should be used for resuscitation, rather than saline. Administration of 0.9% (physiological) saline may result in metabolic acidosis as a result of chloride overload [11]. Additionally, unbuffered 0.9% saline with a pH of 5.0 is irritating and painful when administered subcutaneously [11].
5. Prepare a check list for the operation to ensure you have not forgot anything.
6. You can shave the skin, but we do not recommend it. We have not observed any wound infection in unshaved animals. Our experience has shown that shaving leads to irritation of the skin and thereby to wound complications.
7. To identify the ascendant part of the colon, it is easier to first localize the cecum, which is very big in rats. The ascending colon is the portion of the large intestine, immediately adjacent to the cecum.
8. Wet the colon with the prewarmed Ringer Lactate Solution every 2–3 min.
9. To secure the catheter, try to locate and use the small cuts on the surface of the catheter that were prepared previously.
10. It is not recommended to use a running suture for the skin. The rats try to open the suture sometimes and if the wound is closed with the running suture, the whole cut is opened at once.
11. Rats awake very quickly from the sevoflurane–buprenorphine anesthesia, it normally takes about 5 min until they start to move.
12. Modified Septic Rat Severity Score (SRSS) [12, 13]. The scoring should be performed by a single, blinded researcher (Table 1) to assess the severity of the illness and to protect the animals from unnecessary suffering.
13. The antimicrobials selected for animal studies as well as the timing of the first dose should be chosen very carefully, according to the particular model being used for a given study and the causative pathogen [11].

References

1. Sartelli M, Abu-Zidan FM, Catena F et al (2015) Global validation of the WSES sepsis severity score for patients with complicated intra-abdominal infections: a prospective multicentre study (WISS study). World J Emerg Surg 10:1–8
2. Mulier S, Penninckx F, Verwaest C et al (2003) Factors affecting mortality in generalized postoperative peritonitis: multivariate analysis in 96 patients. World J Surg 27:379–384
3. Martin-Loeches I, Timsit JF, Leone M et al (2019) Clinical controversies in abdominal

sepsis. Insights for critical care settings. J Crit Care 53:53–58

4. Sartelli M, Chichom-Mefire A, Labricciosa FM et al (2017) The management of intra-abdominal infections from a global perspective: 2017 WSES guidelines for management of intra-abdominal infections. World J Emerg Surg 12:1–34
5. Haussner F, Chakraborty S, Halbgebauer R et al (2019) Challenge to the intestinal mucosa during sepsis. Front Immunol 10:891
6. Azevedo LCP (2013) The many facets of sepsis pathophysiology and treatment. Shock 39:1–2
7. Lustig MK, Bac VH, Pavlovic D et al (2007) Colon ascendens stent peritonitis-a model of sepsis adopted to the rat: physiological, microcirculatory and laboratory changes. Shock 28:59–64
8. Zantl N, Uebe A, Neumann B et al (1998) Essential role of gamma interferon in survival of colon ascendens stent peritonitis, a novel murine model of abdominal sepsis. Infect Immun 66:2300–2309
9. Lewis AJ, Seymour CW, Rosengart MR (2016) Current murine models of sepsis. Surg Infect 17:385–393
10. Osuchowski MF, Ayala A, Bahrami S et al (2018) Minimum quality threshold in pre-clinical sepsis studies (MQTiPSS): an international expert consensus initiative for improvement of animal modeling in sepsis. Intensive Care Med Exp 6:26
11. Hellman J, Bahrami S, Boros M et al (2019) Part III: minimum quality threshold in preclinical sepsis studies (MQTiPSS) for fluid resuscitation and antimicrobial therapy endpoints. Shock 51:33–43
12. Herminghaus A, Barthel F, Heinen A et al (2015) Severity of polymicrobial sepsis modulates mitochondrial function in rat liver. Mitochondrion 24:122–128
13. Herminghaus A, Papenbrock H, Eberhardt R et al (2019) Time-related changes in hepatic and colonic mitochondrial oxygen consumption after abdominal infection in rats. Intensive Care Med Exp 7:4

Chapter 3

Induction of Sepsis Via Fibrin Clot Implantation

Sailaja Ghanta, Min-Young Kwon, and Mark A. Perrella

Abstract

Implantation of bacteria embedded in a fibrin clot allows for successful establishment of sepsis in preclinical models. This model allows the investigator to modulate the strain of bacteria as well as the bacterial load delivered. As it allows for a slow release of standardized bacteria, the use of a fibrin clot model may be considered in studying the initial and later phases of sepsis and the host response to infection. Here we describe methods for performing the fibrin clot model of sepsis.

Key words Sepsis, Fibrin clot, Bacteria, Preclinical model, Peritonitis

1 Introduction

Sepsis, a complex and dynamic disease process representing the systemic response to severe infection, remains a major cause of morbidity and mortality across the world [1–3]. The invading microorganism(s) activate resident immune cells that transmigrate to the site of injury resulting in the production of proinflammatory mediators. This proinflammatory response is crucial to trigger phagocytosis and bacterial killing. However, if it continues after the infectious insult has been cleared, it can be harmful to the host and result in collateral organ damage. As a result, following clearance of the infectious insult, resolution of acute inflammation must occur [4]. In addition, if the anti-inflammatory response is excessive, including apoptosis of immune effector cells, then a state of immunoparalysis may occur [5, 6]. This highlights the dynamic nature of sepsis and suggests that improved control of infection and strategies to enhance the host immune response may be important for its treatment.

Numerous clinical and preclinical studies have been attempted to understand the complex pathophysiology of sepsis and study potential treatments [7]. Animal models are used in an effort to create reproducible systems for understanding the pathogenesis of

Wendy E. Walker (ed.), *Sepsis: Methods and Protocols*, Methods in Molecular Biology, vol. 2321,
https://doi.org/10.1007/978-1-0716-1488-4_3, © Springer Science+Business Media, LLC, part of Springer Nature 2021

sepsis [8]. While preclinical studies have shown benefit, most clinical trials of treatments for sepsis have failed to demonstrate efficacy [9, 10]. As a result, it is crucial to keep pursuing preclinical sepsis studies while understanding their limitations as they relate to human disease.

Rodents are a frequently used model of sepsis [10]. They offer great flexibility in that they are readily available in numerous strains including inbred, outbred, and genetically modified. Experimental models in rodents and in other species can be divided into several classes including the use of exogenous toxins such as lipopolysaccharide (LPS), introduction of exogenous bacteria, or alteration of an endogenous protective barrier as in cecal ligation and puncture [7]. The introduction of exogenous bacteria is an ideal model as it allows the investigator to use different doses and strains of bacteria and study the host response. This model can be tailored by altering the route of infection (blood, peritoneal cavity, subcutaneous, lung), frequency of administration (bolus or continuous infusion), bacterial strain, and size of inoculum. All of these parameters can affect progression and outcome. Direct inoculation has been used in a variety of species including mice, rats, and sheep [7, 11]. While it does have some benefits, the acute intravenous injection of live bacteria into animals results in immediate cardiovascular collapse and early death. This does not closely mimic the dynamics of human sepsis [10]. It is more common for patients to have a specific septic focus as opposed to a bolus infusion of bacteria. The fibrin-clot model is a variation of the direct inoculation model and has been proposed as an alternative model of live bacteria injection.

In 1980, Ahrenholz et al. showed that the implantation of bovine fibrin clots containing *E. coli* into the rat peritoneal cavity reduced the 24-h mortality rate from 100% to 0%, compared to bacteria in a similar volume of saline solution. However, the 10-day mortality rate was 90% [12]. The fibrin clot serves as a reservoir for bacterial seeding of the peritoneal cavity, other organs, and subsequently the bloodstream, similar to what is observed in human sepsis [11]. It consists of entrapping a set dose of bacteria into a fibrin clot and performing a midline laparotomy to implant the fibrin clot into the peritoneal cavity [12]. Fibrin delays the systemic absorption of the entrapped bacteria and promotes the development of an abscess, which serves as a more local septic focus that causes a peritonitis and ultimately a bloodstream infection [10, 12]. The benefits of this model are that it allows a slower release of bacteria and a more sustained infection.

The fibrin clot model is thought to be highly reproducible. It has been described in both small and large mammals [13]. In addition, it displays many of the features of human sepsis including insidious onset, hyperdynamic cardiovascular state, reversible left ventricular dilatation with impaired systolic performance, and a significant but delayed mortality rate [13, 14]. It has also been shown to produce inflammatory changes in the liver and spleen, as well as positive blood and peritoneal cultures [15–17]. This model is highly reproducible as bacterial quantities can be controlled [13]. Most of the studies using fibrin clot models of sepsis have been done using single-organism cultures. However, mixed cultures might more accurately mimic the gastrointestinal flora in peritonitis.

Here, we describe methods for creating the fibrin clot and implanting it in the peritoneum for development of sepsis in mice. In this protocol, we use *E. coli*. However, any bacteria can be used. In addition, we describe common methods of analyzing sepsis in this model including survival, organ injury, and bacterial burden.

2 Materials

2.1 Clot Preparation

1. 6.7% fibrinogen (weight/volume, sterile filtered) (*see* **Note 1**).
2. Thrombin 2 U/mL.
3. Glycerol bacterial stock—*E. coli* or other bacteria (*see* **Note 2**).
4. Sterile phosphate buffered saline (PBS).
5. LB media.
6. Spectrophotometer and cuvettes.
7. LB plates.
8. 15 mL tubes.
9. Glycerol.
10. Pipettes and tips.
11. Cover of 6-well plate.

2.2 Surgery

1. C57BL/6 mice aged 6–10 weeks, male or female (*see* **Note 3**).
2. Ketamine (100 mg/kg body weight; *i.p.*).
3. Xylazine (10 mg/kg body weight; *i.p.*).
4. 23 gauge needles.
5. 70% ethanol.
6. Razor blades or electric trimmer.
7. Surgical instruments (scalpel, dissection scissors, needle driver, forceps without teeth, straight surgical scissors).
8. Platform for animal surgery.
9. Heater.
10. 6–0 surgical suture (absorbable vicryl and nonabsorbable Prolene)
11. Lamp.

12. 1 mL syringe.
13. Sterile gloves, mask, and drapes.
14. Buprenorphine 0.05–0.1 mg/kg.
15. Betadine.
16. 0.9% saline solution.

3 Methods

Before commencing, it is essential to make sure all of the surgical procedures have been approved by the institutional and national ethical committees. All procedures should be performed aseptically.

3.1 Clot Preparation

1. To create a culture of *E. coli*, place 10 mL of LB media into a 15 mL conical tube. Using sterile technique and a flame, scrape the frozen glycerol stock with a pipette tip to obtain a small amount of material and inoculate this into the LB media.
2. Incubate for 16 h on a rotary shaker (200 rpm) at 37 °C, typically overnight.
3. After 16 h, take 1 mL of the bacteria and regrow it in 9 mL of LB media for 3 h at 37 °C on a rotary shaker to ensure cells are in a state of growth.
4. After 3 h, pellet the bacteria by centrifugation at 3500 rpm (2400 × *g*) for 10 min. Wash with 10 mL PBS twice and resuspend in 1 mL PBS.
5. Use a spectrophotometer to measure the optical density at 600 nm (OD_{600}) and calculate the concentration (*see* **Note 4**).
6. Adjust the suspension with PBS to yield a concentration of 55 x 10^9 CFU/mL. 40 μL of this will be used to make each clot, corresponding to 2.2 × 10^9 CFU/clot (*see* **Note 5**).
7. The actual dose of the clot should be assessed retrospectively via a colony-forming assay (*see* **Note 6**).
8. Prepare the clot by aliquoting 130 μL of 6.7% Fibrinogen (weight/volume, sterile filtered). Then add 190 μL of PBS and bacteria. We usually add 40 μL of bacteria as noted above and 150 μL of PBS (*see* **Note** 7).
9. To form a base for the clot, place a drop of glycerol on the cover of a 6 well plate (*see* **Note 8**).
10. Add 8 μL of thrombin (2 U/mL) to the tube with Fibrinogen and PBS/bacteria.
11. Quickly mix with a pipet while avoiding bubbles and apply the total mixture to the glycerol drop to form a circle (see **Note 9** and **10**).

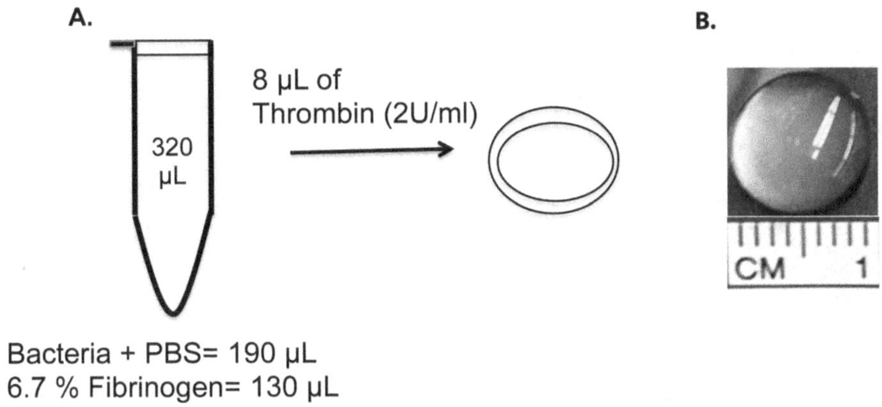

Fig. 1 Creation of the Fibrin Clot. (**a**) In this panel, the components of the fibrin clot including bacteria, phosphate buffered saline (PBS), fibrinogen, and thrombin are demonstrated. Once thrombin is added, the clot is allowed to remain at room temperature for 10 min. (**b**) In this panel, a representative picture and size of the formed clot is shown

12. Wait 10–15 min for a clot to form (*see* Fig. 1).
13. Vehicle clots should contain additional PBS and no bacteria to make the same clot volume.

3.2 Surgery

1. Weigh the animals.
2. Anesthetize the mice with a combination of xylazine (10 mg/kg body weight, *i.p*) and ketamine (100 mg/kg body weight, *i.p.*).
3. Shave the abdomen with the razor blade or electric trimmer, being careful not to cut the skin.
4. Disinfect the surgical area with a betadine solution followed by 70% alcohol. Each is done in triplicate.
5. Place the mice on a surgical platform and cover with surgical drapes.
6. Using a dissection scissors or scalpel, make a 1.5 cm midline incision in the skin being careful not to puncture the peritoneal cavity.
7. Use a small scissors, make another ~1 cm midline incision in the peritoneal membrane, being careful not to puncture abdominal organs or the bowel (*see* Fig. 2a).
8. Using a sterile forceps without teeth, grasp the clot or the vehicle clot (for the sham animals) (*see* Fig. 2b and **Note 10**).
9. Place the clot within the peritoneal cavity in the right upper quadrant (*see* Fig. 2c and **Note 11**).
10. The peritoneum and skin are closed in 2 layers. The peritoneum is closed with absorbable 6–0 vicryl sutures. The skin is closed with non-absorbable 6–0 prolene sutures with a simple running stitch.

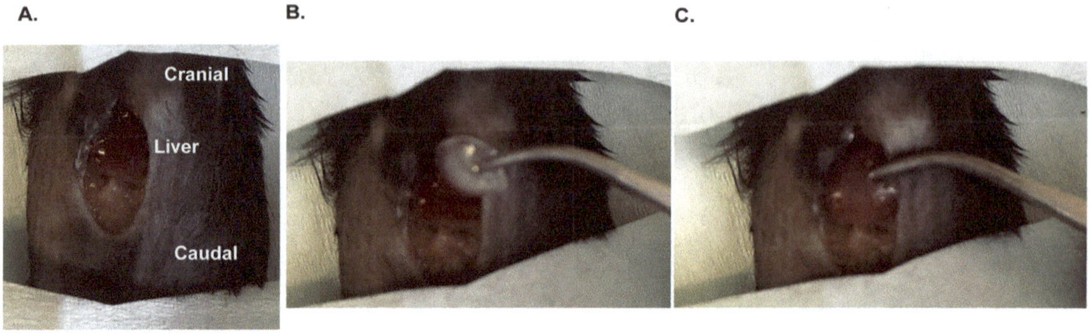

Fig. 2 Insertion of the Fibrin Clot. This figure demonstrates insertion of the created fibrin clot into the peritoneal cavity. (**a**) In this panel, the peritoneum is opened, and the right upper quadrant is exposed. (**b**) In this panel, forceps are used to grasp the clot and slide it into the peritoneal cavity as shown. (**c**) In this panel, the clot is shown inserted into the peritoneum in the right upper quadrant, shown here inserted above the liver. The peritoneum and skin are then closed in two layers

11. Inject each mouse with 1 mL of sterile 0.9% saline subcutaneously for resuscitation.
12. Place the mouse on a heating pad to recover.
13. After the mouse recovers from anesthesia (typically 30 min to 1 h) place the animal back in its cage with food and water.
14. Give the mice post-operative analgesics (buprenorphine 0.05–0.1 mg/kg subcutaneously every 12 h × 48 h) (*see* **Note 12**).

3.3 Assessment of Sepsis

Once the fibrin clot model of sepsis is initiated, it can be used to test therapeutics for sepsis. In addition, by using genetically modified mice, it can be used to study the pathophysiology of sepsis. Several endpoints can be used to evaluate the degree of sepsis. The more common ones include:

1. Survival: Following surgery and recovery, mice are monitored for up to 7 days. Most mortality in the fibrin clot model occurs between 24 h and 3 days. We monitor the mice closely, and place food and water in the bottom of the cage for easy access. Humane endpoint criteria should be determined to identify mice that are in pain or moribund. Mice that are moribund or in pain (ruffled fur, hunched position, lack of activity, movement, or eating/drinking, weight loss, respiratory distress, etc.), should be euthanized by methods approved by the IACUC.
2. Blood pressure can be monitored by placement of intra-arterial catheters in the carotid arteries [15].
3. Bacterial load in the blood and peritoneal fluid: 24 h after fibrin clot placement, peritoneal lavage can be performed with 7 mL of PBS. The peritoneal fluid can then be assessed for colony

forming units (CFUs) of bacteria. Serial dilutions of the peritoneal fluid are made, plated on LB agar plates, and then incubated overnight.

4. Blood can be obtained via cardiac puncture and assessed for colony forming units (CFUs) of bacteria in a similar fashion [18]. While we typically perform this at 24 h, any time point can be used to assess bacterial load.
5. Organ injury: 24 h after fibrin clot, organs including the spleen, liver, kidney, bowel, and lungs can be harvested and fixed in Formalin or Methyl Carnoy (fixative depends on specifics of antibody). Tissue sections can be assessed for morphology and stained with antibodies that detect neutrophils (LY6G) and macrophages (F4/80) to assess organ inflammation [15]. While we typically perform this at 24 h, any time point can be used to assess organ injury.

4 Notes

1. A stock fibrinogen concentration of 6.7% yields a final fibrinogen concentration of 2.6% (the standard concentration described in this protocol). We and others have used fibrinogen concentrations ranging from 0.5–2.6% [12, 15]. These concentrations create a clot that is digestible but still solid enough for manipulation. The severity of sepsis is determined by the bacterial concentration and size of the clot, and is not affected by the concentration of fibrinogen [12].
2. For our glycerol stocks of bacteria, *E. coli* was identified by culturing contents of the ileum (Channing Laboratory, Brigham and Women's Hospital) and stored in LB media containing 20% glycerol at −80 °C. In this protocol we describe an *E. coli* fibrin clot, but any bacteria can be used for fibrin clot creation depending on the intended study. We have used *E. faecalis* as a gram-positive bacteria in fibrin clots.
3. This protocol employs the C57BL/6 strain, but any strain of mouse can be used. As described in numerous sepsis models, several parameters can affect sepsis severity. These include the age, gender, and strain of the mice. Several studies have described female mice as being more resistant to sepsis than their male counterparts [19]. In addition, the time of day the surgery is performed can also affect outcomes [11]. It is best to keep all of the above parameters as consistent as possible. In addition, it is also suggested to repeat the fibrin clot in different strains of mice or genders to ensure applicability to clinical disease.

4. Concentration should be estimated by the absorbance. Prior to starting, it is essential to create a bacterial standard curve to correlate CFU/mL with OD measurements as these can vary. To create your standard curve, during growth phase, take samples of the bacterial culture in regular intervals. Perform OD measurements as well as plating of diluted samples on LB agar plates. After overnight incubation count the colonies and calculate CFU/mL. Plot OD and CFU/mL values to create your standard curve. We found that an OD_{600} of 1 gave us 55×10^9 CFU/mL. This number can vary based on numerous factors including the bacteria, the spectrophotometer, and the cuvette used so it is necessary to generate your own standard curve.
5. We chose this concentration of bacteria to cause ~40–50% lethality by 7 days. However, bacterial counts and the resulting sepsis severity can be titrated based on the strain of bacteria, animal susceptibility, and desired endpoints.
6. In addition, the bacterial concentration in the suspension and the clot should be confirmed retrospectively in each experiment by plating on LB agar and counting colony-forming units. This can be accomplished by making 5 serial tenfold dilutions of the culture. 100 μL of the 10^{-4} and 10^{-5} dilutions should be spread onto LB agar plates and incubated overnight at 37 °C. Then, count the number of colonies to determine the actual administered dose.
7. If desired, a master mix of bacteria, PBS, and Fibrinogen can be made for the total number of clots and then divided into aliquots of 320 μL. A drop of glycerol can be placed on the lid of a 6-well plate and the clot placed over it, to prevent the clot from sticking to the plate. This will make it easier to handle.
8. Be ready with all of your supplies (pipet to mix, glycerol drop and 6 well plate lid) because once thrombin is added, the clot is rapidly formed. Once you add thrombin, quickly mix and then pipet your clot onto the glycerol drop.
9. If bubbles form in the clot, you can use a needle to puncture them.
10. We have found that forceps without teeth are best to handle the clot firmly without breaking it.
11. The clot can be placed anywhere in the peritoneum. We use the right upper quadrant for consistency.
12. In order to study the pathophysiology of single organism sepsis and its mechanisms in this fibrin clot model, we do not give antibiotics. However, to more accurately mimic the clinical scenario, antibiotics can be given.

Acknowledgments

This work was supported by the National Institutes of Health grants K08GM126313 (to Dr. Ghanta) and R01GM118456 (to Dr. Perrella).

References

1. Angus DC, Linde-Zwirble WT, Lidicker J et al (2001) Epidemiology of severe sepsis in the United States: analysis of incidence, outcome, and associated costs of care. Crit Care Med 29:1303–1310
2. Annane D, Bellissant E, Cavaillon JM (2005) Septic shock. Lancet 35:63–78
3. Martin GS, Mannino DM, Eaton S et al (2003) The epidemiology of sepsis in the United States from 1979 through 2000. N Engl J Med 348:1546–1554
4. Pinsky MR (2001) Sepsis: a pro- and anti-inflammatory disequilibrium syndrome. Contrib Nephrol 132:354–366
5. Hotchkiss RS, Nicholson DW (2006) Apoptosis and caspases regulate death and inflammation in sepsis. Nat Rev Immunol 6:813–821
6. Weber SU, Schewe JC, Lehmann LE et al (2008) Induction of Bim and Bid gene expression during accelerated apoptosis in severe sepsis. Crit Care Med 12:R128
7. Stortz JA, Raymond SL, Mira JC et al (2017) Murine models of sepsis and trauma: can we bridge the gap? ILAR J 58:90–105
8. Buras JA, Holzmann B, Sitkovsky M (2005) Animal models of sepsis: setting the stage. Nat Rev Drug Discov 4:854–865
9. Piper RD, Cook DJ, Bone RC et al (1996) Introducing critical appraisal to studies of animal models investigating novel therapies in sepsis. Crit Care Med 24:2059–2070
10. Poli-de-Figueiredo LF, Garrido AG, Nakagawa N et al (2008) Experimental models of sepsis and their clinical relevance. Shock 30(Suppl 1):53–59
11. Lewis AJ, Seymour CW, Rosengart MR (2016) Current murine models of Sepsis. Surg Infect 17:385–393
12. Ahrenholz DH, Simmons RL (1980) Fibrin in peritonitis. I. Beneficial and adverse effects of fibrin in experimental E. coli peritonitis. Surgery 88:41–47
13. Mathiak G, Szewczyk D, Abdullah F et al (2000) An improved clinically relevant sepsis model in the conscious rat. Crit Care Med 28:1947–1952
14. Natanson C, Fink MP, Ballantyne HK et al (1986) Gram-negative bacteremia produces both severe systolic and diastolic cardiac dysfunction in a canine model that simulates human septic shock. J Clin Invest 78:259–270
15. Baron RM, Kwon MY, Castano AP et al (2018) Frontline science: targeted expression of a dominant-negative high mobility group A1 transgene improves outcome in sepsis. J Leukoc Biol 104:677–689
16. Chung SW, Liu X, Macias AA et al (2008) Heme oxygenase-1-derived carbon monoxide enhances the host defense response to microbial sepsis in mice. J Clin Invest 118:239–247
17. Toky V, Sharma S, Arora BB et al (2003) Establishment of a sepsis model following implantation of Klebsiella pneumoniae-infected fibrin clot into the peritoneal cavity of mice. Folia Microbiol 48:665–669
18. Tsoyi K, Hall SR, Dalli J et al (2016) Carbon monoxide improves efficacy of mesenchymal stromal cells during sepsis by production of specialized proresolving lipid mediators. Crit Care Med 44(12):e1236–e1245
19. Zellweger R, Wichmann MW, Ayala A et al (1997) Females in proestrus state maintain splenic immune functions and tolerate sepsis better than males. Crit Care Med 25:106–110

Chapter 4

Cecal Slurry Injection in Neonatal and Adult Mice

Jaimar C. Rincon, Philip A. Efron, Lyle L. Moldawer, and Shawn D. Larson

Abstract

Studying the pathophysiology of sepsis still requires animal models, and the mouse remains the most commonly used species. Here we discuss the "cecal slurry" (CS) model of polymicrobial, peritoneal sepsis and compare and contrast it to other commonly used methods. Among the different murine models of sepsis, cecal ligation and puncture (CLP), and not the CS, is often considered the "gold standard" to induce polymicrobial sepsis in laboratory animals. CLP is a well-described model involving a simple surgical procedure that closely mimics the clinical course of intra-abdominal sepsis. However, CLP may not be an option for experiments involving newborn pups, where the cecum is indistinguishable from small bowel, where differences in microbiome content may affect the experiment, or where surgical procedures/ anesthesia exposure needs to be limited. An important alternative method is the CS model, involving the intraperitoneal injection of cecal contents from a donor animal into the peritoneal cavity of a recipient animal to induce polymicrobial sepsis. Furthermore, CS is an effective alternative model of intraperitoneal polymicrobial sepsis in adult mice and can now be considered the "gold standard" for experiments in neonatal mice.

Key words Infection, Cecal slurry, Sepsis, Polymicrobial sepsis, Neonatal sepsis, Murine sepsis

1 Introduction

Despite significant advances in critical care medicine, sepsis represents a major public health challenge worldwide [1, 2]. Although the global epidemiological burden of sepsis is difficult to completely ascertain, the World Health Organization (WHO) estimates that more than 30 million people worldwide are diagnosed with sepsis leading to 6 million deaths per year [3]. Sepsis is particularly lethal at the extremes of age (i.e., during the neonatal period and in older adults). Neonates, particularly preterm (<37 weeks gestation) and low-birth weight (<2500 g) infants are left vulnerable to sepsis due to multiple defects in host protective immunity [4, 5]. The incidence of sepsis in neonates and children is estimated to account for 3 million and 1.2 million cases, respectively [6].

Wendy E. Walker (ed.), *Sepsis: Methods and Protocols*, Methods in Molecular Biology, vol. 2321,
https://doi.org/10.1007/978-1-0716-1488-4_4,

Importantly, the management of sepsis and its long-term morbidity incur high costs, making sepsis the most expensive medical condition treated in the US healthcare system [7]. As sepsis places a substantial patient and financial burden on healthcare systems, it is imperative that we understand its complex pathophysiology so that novel therapeutic targets and effective treatment strategies can be developed. Currently, the use of animal models for the study of sepsis remains an important component of the translational research.

Promising therapeutic strategies to treat sepsis have been developed in preclinical studies using murine models (mouse; *Mus musculus*) [8]. While it has been argued that animal models do not fully recapitulate the human condition in sepsis [9, 10], they remain an important translational research tool [11, 12]. Challenges with using murine models for sepsis research include their reduced responsiveness to microbial products, their different blood leukocyte profile, and the inability to support organ function during critical illness [13]. Despite these limitations, murine models of sepsis have been helpful at better understanding individual molecular mechanisms underlying the complex pathophysiology of sepsis. The mouse's uniform genetic background, availability of various strains, high fecundity, ease of handling, and familiarity within the research community have made it the model of choice [11].

Historically, murine sepsis models relied on the intravenous or intraperitoneal administration of either endotoxin or individual strains of live bacteria [14, 15]. These models produced massive non-physiologic increases in most inflammatory mediators associated with hypovolemia and cardiac failure and induced rapid death in these animals [13]. Although these models continue to be used, they are not frequently intended to replicate human sepsis in any but the most unusual circumstances.

More recently, different animal models of intraperitoneal sepsis have been developed, in most of which systemic inflammation is initiated by a local intra-abdominal infection. Among the murine models of intraperitoneal sepsis, cecal ligation and puncture (CLP), is one of the most frequently used and generally accepted as the "gold standard." CLP is a well-understood surgical procedure that is thought to closely resemble the progression and characteristics of the complexity of human sepsis [16]. Despite its widespread use, the CLP model often does not satisfy one critical aspect of sepsis management: source control [17]. Despite antibiotic treatment, the presence of a devitalized cecum assures failure of any antibiotic intervention to completely resolve the infection, and involves both abscess formation and chronic inflammation. Complete recovery in this model cannot be assured.

There are a number of alternatives to a CLP, including a cecal stent (colon ascendens stent peritonitis, CASP), inoculation of the peritoneum with standardized amounts of bacteria in fibrin clots or extracellular matrix-based hydrogel, or administration of fecal contents [13, 18]. Cecal stents have the advantage of producing persistent peritonitis, and does not involve a devitalized cecum [19]. However, like the CLP model, resolution of the infectious process requires surgical intervention to obtain source control. An infected blood clot eliminates the need for source control, but is technically challenging and difficult to reproduce. Some investigators have used it for single species infections, but this method has not been widely embraced in the research community.

An alternative method to CLP has been proposed involving the intraperitoneal administration of the cecal contents of a euthanized animal, into another animal to induce polymicrobial sepsis [11, 20–22]. This offers a number of theoretical and technical advantages. Cecal contents contain not only microbes but also particulate matter that assists with bacterial colonization of the peritoneum. In an untreated animal, bacterial colonies can be recovered transiently from the blood, and persistently in the peritoneum and visceral organs. In contrast to CLP, appropriate antibiotic interventions can completely eliminate mortality and bacterial colonization, and source control is not required.

The cecal slurry (CS) model has become the preferred method by investigators studying sepsis in neonatal mice, as it is an infectious model characterized by bacterial colonization, systemic inflammation, and dose-dependent mortality without surgery, which may be lethal in murine neonatal pups [20, 23–28]. This model is designed to more closely replicate the human neonatal condition of necrotizing enterocolitis (NEC), a devastating condition predominantly in preterm neonates. CS has been widely used by numerous laboratories and is now considered the "gold standard" model for murine neonatal sepsis research. Apart from avoiding surgery in newborn mice, CS-induced sepsis can be utilized in a large number of animals from a single CS donor and mimics polymicrobial sepsis often seen in neonatal bowel perforation. Additionally, the technique is acceptably reproducible, involves no surgical manipulation and is easy to perform by researchers. The latter is particularly important as CLP or cecal stent in a neonate would require a level of surgical expertise under a microscope or magnifying glass not present in most laboratories. Additionally, these methods would add significant time constraints to be able to efficiently conduct these experiments in a humane and successful manner under anesthesia.

It is important to note, however, that any peritoneal sepsis model, whether CLP, cecal stent, or CS produces a unique response to severe infection that is both dose dependent and in which the severity of the sepsis can be modified. For example, we evaluated the inflammatory cytokine and genomic response in juvenile mice to a similarly lethal CLP and CS, and found that not only were the patterns of inflammatory cytokine responses different between the two models, but the genomic changes from blood leukocytes were dramatically different, but reproducible [11]. Despite a comparable mortality, signaling pathways activated by CLP and CS are different. While the CS model results in early inflammatory response, CLP is characterized by down regulation of T cell activation and increased expression of immune suppressive pathways [11].

2 Materials

2.1 Animals

Polymicrobial sepsis induced by CS can be administered to neonatal mice (5–7 days-old), young adult (6–12 weeks) or elderly (>18 weeks) mice, housed under specific pathogen-free conditions (*see* **Note 1**).

2.2 Cecal Slurry Preparation

1. CS donors: 6–12-week-old female mice to induce neonatal sepsis and age matched CS donors to induce sepsis in young adult mice or aged mice.
2. CO_2 chamber.
3. Sterile straight-edge scissors, forceps, and spatula (*see* **Note 2**).
4. Sterile 5% dextrose.
5. 70% ethanol.
6. Sterile petri dish.
7. 15 mL conical polypropylene centrifuge tube.
8. Scale.

2.3 Cecal Slurry Injection

1. Conventional insulin syringes (1 mL).
2. Needles 25G × 5/8″, regular bevel.
3. Scale.
4. Heating pad.

3 Methods

3.1 Cecal Slurry Preparation

1. Euthanize the CS donor mouse by hypoxia, exposing it to 10–30% CO_2 inhalation following AVMA Guidelines for euthanasia of animals until respiration and heartbeat have ceased. An adjunctive physical method (e.g., exsanguination, cervical dislocation) should be used to ensure death in an already euthanized animal (*see* **Note 3**).
2. Fix the mouse to a polystyrene foam board and spray the lower abdomen with 70% ethanol. Using a straightedge scissor and forceps, open the skin and the abdominal musculature to

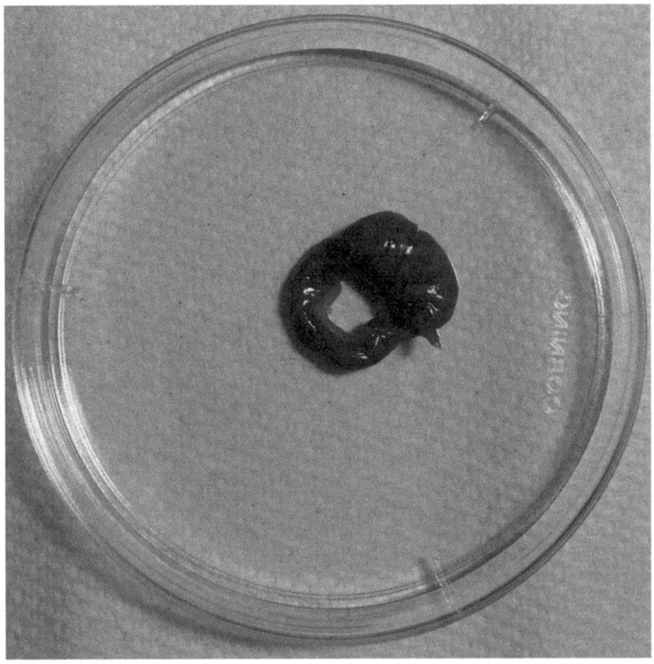

Fig. 1 Excised cecum aseptically collected from a mouse donor. Following euthanasia of the CS donor mouse, open the skin and localize the cecum at the junction of the ileum and the proximal colon. Excise the cecum and place it on a sterile disposable petri dish

localize the cecum switching to clean instruments in between. Once the abdominal cavity is open, localize the cecum (comma-shaped blind sac-like organ) at the junction of the ileum and the proximal colon. Excise the cecum and place it on a sterile disposable petri dish (Fig. 1).

3. Using sterile forceps and a spatula, extrude the cecal contents out of the epithelium into a preweighted 15 mL conical polypropylene centrifuge tube.
4. Weigh the cecal contents and prepare 80–100 mg/mL CS with 5% dextrose.
5. Vortex the CS solution for 20–30 s to make a homogeneous suspension (Fig. 2).
6. To ensure maximal effectiveness of the CS, it is recommended to use the solution within 2 h of preparation (*see* **Note 4**).

3.2 Cecal Slurry Injection in Neonatal Mice

The age of a neonatal mouse can be determined by their physical characteristics during the first week of life and varies according to the strain. To better record the day of birth, pregnant dams may be monitored daily 2–5 days prior delivery. For experiments with neonatal mice, mixed gender pups aged 5–7 days are challenged with CS to induce polymicrobial sepsis.

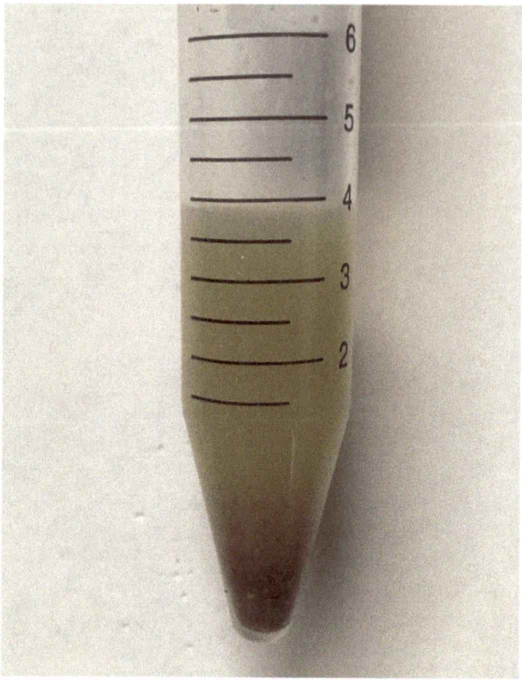

Fig. 2 CS solution. Extrude the cecal contents out of the epithelium onto a preweighted conical polypropylene centrifuge tube, and prepare 80–100 mg/mL CS with 5% dextrose

1. Separate the litter from the dam by placing them in a new cage. Leave part of the nest on the cage to reduce anxiety to the dam and avoid behavioral changes (e.g., neglect of the pups; *see* **Note 5**).
2. If pups are to receive pretreatment interventions prior to sepsis mark the tail using a permanent marker (*see* **Note 6**).
3. Weigh each pup to calculate the weight-adjusted dose of the CS solution (80 mg/mL) to induce the required lethality (*see* **Note 7**).
4. Depending on the number of litters to be challenged by CS, fill the insulin syringes with up to 200–300 μL of CS solution.
5. Using the thumb and index fingers, secure the pup by the back of the neck and the lower back.
6. Insert the needle subcutaneously with the bevel facing up in the lower abdomen (Fig. 3). Inject the weight-adjusted dose of CS (1.1–1.3 mg/g body weight), waiting 10 s to ensure that the correct CS dose is delivered. Carefully withdraw the needle to avoid spillage or injury to the pup. Placing the injected pup on a paper towel will help to determine any CS leakage or bleeding following injection (*see* **Note 8**).

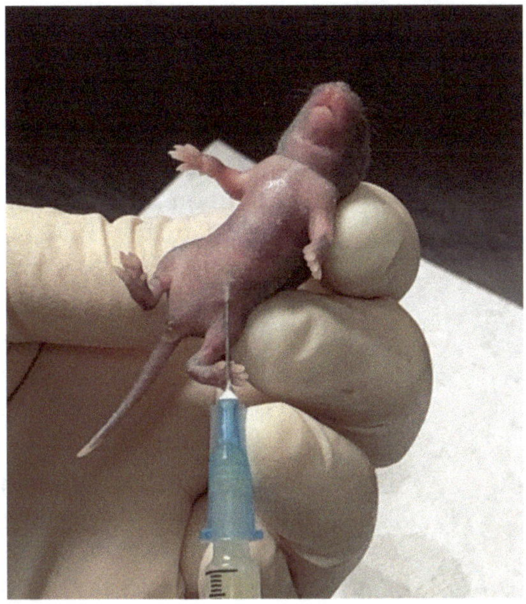

Fig. 3 Injection of CS into a 6-day-old B6 neonatal mouse. Secure the pup by the back of the neck and the lower back and inject the weight-adjusted dose of CS (1.1–1.3 mg/g body weight), waiting 10 s to ensure that the correct CS dose is delivered

7. Once all pups have been injected and monitored, return the litter to the cage with the dam.
8. Following CS injection, neonatal mice should be monitored following the local Institutional Animal Care and Use Committee approved protocol (*see* **Note 9**). In general, following standard CS dosing, monitor all injected neonatal mice for the next 2 h to identify any injection-related complications; after that, neonatal mice should be monitored at a minimum every 12 h to identify post-sepsis morbidities (scattering, absence of milk spot). Scattering is when the mother abandons the pup and it is physically separated from the mother and other pups. Scattering, poor feeding (evidenced by the absence of a milk spot), and/or dam neglect often indicates that the pup is moribund and will not survive, and therefore should be euthanized for animal welfare concerns per institutional protocol (these pups are considered non-survivors for the experiment) (*see* **Note 5**).

3.3 Cecal Slurry Injection in Adult Mice

Although CLP is the most widely used model of intra-abdominal sepsis in adult mice, it has been shown that variability in the surgical technique may play a significant role in outcome [29]. The CS model of sepsis is particularly useful to study early inflammatory components of sepsis or as an alternative model of sepsis for those animals with different cecum size, bacterial flora or animals with deficiency in wound-healing capability (increased sensitivity to surgery).

1. Weigh each mouse to calculate the weight-adjusted dose of the CS solution (1.1–1.3 mg/g body weight) to induce the required lethality (LD_{40-70}) (*see* **Note 7**).
2. Remove the animal from the cage and gently restrain in the head-down position. Tilt the mouse (abdomen side up) with its head slightly toward the ground so that its head is lower than its hind end (in this position any accidental puncture of abdominal organs will be minimized). Prep the site of injection with 70% ethanol.
3. Inject the weight-adjusted dose of CS in the mouse's lower right quadrant of the abdomen (just above the level of the hip). The needle should be inserted with bevel facing "up" at a 30–40° angle to horizontal, to avoid damage to the bladder and other abdominal organs. Pull back on the plunger to ensure negative pressure before injecting the CS, and insert the entire bevel within the abdominal cavity.
4. Place the injected mouse in a new cage and monitor closely, following the local Institutional Animal Care and Use Committee approved protocol. In adult mice, lethargy, piloerection, hypothermia and hunching posture are examples of primary signs of sepsis. Moribund animals (prostrate, unresponsive, and unable to right itself) should be euthanized per local protocol to avoid spontaneous death [17] (*see* **Note 9**).

3.4 Comments

Treatment of sepsis is a major issue to the US healthcare system due to the high morbidity and mortality, costing billions of dollars annually. Although there are limitations in the animal models of sepsis, they are required for translational research. The CLP murine model of polymicrobial sepsis, initially developed by Chaudry et al. [16], has been widely used over the past years as the "gold standard" model to study the pathophysiology of sepsis in animals. In PubMed alone, more than 3000 articles are found using the key words "cecal ligation and puncture and sepsis." This is clearly due to the advantages that this model offers to the scientific community focused on the study of host immune reactions to sepsis (a simple experimental procedure that has been extensively described).

Although some variations have been made to this model through the years to increase the severity of peritonitis (needle size, more than one puncture), another two models developed to improve CLP reproducibility are the colon ascendens stent peritonitis (CASP) and the implantation models using bacteria-impregnated fibrin clots [30, 31]. These two models of peritonitis are different to CLP in terms of pathogenesis and disease progression. While the CLP model is useful to study peritoneal abscess formation accompanied by local peritoneal infection, inflammation and infiltration, the implantation of fibrin clot and CASP models are better options to study enterobacterial peritoneal infection accompanied by bacteremia, SIRS, and sepsis progression without the presence of devitalized tissue [32, 33].

Although these surgical models of sepsis (CLP, fibrin clot and CASP) cover different aspects of sepsis, they do have some important disadvantages for research including experiments using animals with gastrointestinal pathologies, wound-healing deficiencies, different cecum size or shape, and newborn animals. The endotoxemia and bacteria administration models and the intraperitoneal injection of cecal content from a donor rodent (CS model) are nonsurgical experimental models of sepsis developed as an alternative to the surgical procedures mentioned before (Table 1).

The CS model initially developed to induce sepsis in adult pigs [34], was undertaken to validate it as a model of neonatal sepsis in mice by Wynn et al. [20]. This model of sepsis is characterized by bacterial colonization, systemic inflammation, inflammatory-related splenic alterations, and a dose-dependent mortality [20]. The main advantage of the CS model in neonatal mice, is that it can be utilized in pups as young as 5-days old in place of a surgical model of sepsis (CLP, CASP, fibrin clot). Surgical models of sepsis are not possible to perform in the neonatal mouse due to their body size and the incomplete intestinal development as well as the need for prolonged anesthesia and highly technical surgical expertise). Additionally, CS can be performed with less risk of maternal neglect (a common behavior in laboratory rodents) [20, 35]. Other nonsurgical models of sepsis (LPS or bacteria injection) have inherent limitations. Although LPS is a major component of the outer membrane of gram-negative bacteria, it is a single molecule of the complex pathogen associated molecular patterns (PAMPs) of these organisms. Conversely, this model neglects the host–pathogen interactions of gram-positive bacteria as well as polymicrobial sepsis. Furthermore, bolus administration of LPS has been described as an intoxication model rather than sepsis [17, 36]. Similarly, intraperitoneal or intravenous bacteria injection to induce sepsis in mice have important limitations [10, 13]. Large inocula are needed due to the inability of bacteria to replicate in vivo, and in order to overcome murine host defense [14, 37]. In addition, like endotoxicosis, the host early inflammatory response is massively elevated, several logs higher than seen in septic humans.

Table 1
Experimental models of sepsis

	Nonsurgical models of sepsis		Surgical models of sepsis		
	Bacteria or endotoxin (LPS) injection	**Cecal slurry (CS) injection**	**Cecal ligation and puncture (CLP)**	**Bacteria and fibrin clot implantation**	**Colon ascendens stent peritonitis (CASP)**
Description	Intraperitoneal or intravenous injection of bacteria or purified LPS	Intraperitoneal injection of cecal contents from a donor	Ligation of a portion of the cecum followed by cecal colostomies via one or more needle punctures	Standardized amounts of bacteria in fibrin clots, implanted into the peritoneal cavity	Insertion of a stent into the ascending colon to allow the stool to continuously flow from the colon into the peritoneal cavity
Pros	– Easy to procedure – Replicates the physiology of severe sepsis – Highly reproducible – Dose of bacteria or endotoxin can be standardized	– High reproducibility – No surgical tissue trauma or ischemic tissue – Advantageous in experiments that require large number of animals – Can be performed in neonatal animals with less risk of being neglected by the mother – Standardized polymicrobial inoculum (weight-based intraperitoneal fecal injection)	– Easy procedure – Partly reflects human disease – Extensively studied – Sepsis severity can be adjusted based upon the length of the cecum ligated and the size and number of punctures performed – Better to study peritoneal abscess formation, local inflammation and infiltration	– Fibrin delays systemic infection – Replicates features of sepsis in humans – Low early mortality – Highly reproducible	– Mimics human sepsis better than CLP – Better to study bacteremia, SIRS and sepsis – Sepsis severity can be adjusted by altering the size of the catheter
Cons	– Transient inflammatory response – Intoxication model (septic shock) – Host response limited to gram-negative organisms (LPS) or bacteria injected	– High variability (lethality of the cecal preparation varies between donor batches)	– Surgical trauma – Time-consuming – Bacterial dissemination in the peritoneum (more localized inflammation) – High variability – Impossible to control rate and amount of fecal material released	– Surgical trauma – The use of a single organism in the fibrin clot reduces the clinical relevance	– Time-consuming – poorly characterized – High variability

The lack of clinically predictive animal models remain as an important barrier to develop effective therapeutic strategies to treat sepsis. Antibiotic therapy intervention has been included to mimic the clinical condition and can substantially alter the model of sepsis [38–41]; however, the timing of therapeutic interventions, dosage and antibiotic class are inconsistent with the antibiotic therapy used in humans. Further, while experimental models of sepsis use a single strain of inbred mice with a single infection, in humans there is a heterogeneous population (genetics, age, disease severity). The CS is an acute model of sepsis developed to induce a systemic inflammatory response in neonatal mice; antibiotic treatment in neonatal mice may have a greater impact on the immune system than adult exposure and represents a technical challenge. Additionally, the impact of antibiotics on inflammation in murine models of sepsis might add additional variables that could hinder the molecular mechanisms that mediate poor sepsis outcomes; therefore, it needs to be further investigated.

Although the clinical relevance of murine models of sepsis are still a topic of debate due to the substantial differences to human responses to sepsis [9, 10], there are biological similarities that may be highlighted as they represent a better option to study the complex pathophysiology of sepsis. As the CS model may be performed in adult and neonatal mice, this model of polymicrobial sepsis is a good option not only to compare neonatal and adult sepsis but also to develop novel therapeutic agents in sepsis especially designed for these two different populations [23, 24, 42, 43].

4 Notes

1. Genetics of the mice is important to consider. Although the background of inbred mice is genetically uniform, inbred strains show important variations in their immune response due to genetic mutations and polymorphisms [44]. For example, caspase-1/11, IL-18, and PD-1 (B6 background) null mice are more resistant to polymicrobial sepsis induced by CS compared to wild type mice [23, 28, 45, 46]. Conversely, CS administration does not affect survival on ASC, NLRP3, TRIF, MyD88, or RAG1-null neonatal mice compared to WT (B6) [24, 45, 47], as well as in C3H/HeJ mice [24].
2. All surgical instruments (stainless-steel forceps, straight-edge scissors, and spatula) should be heat-sterilized as would be in any surgical procedure. However, because harvesting of the fecal contents is a nonsurvival procedure and considered "dirty," instruments may be sterilized using a preheated hot bead sterilizer for 10–20 s (**Caution:** do not leave instruments

longer than 20 s as they will become too hot to handle safely and will cause thermal injury).

3. Following CO_2 exposure, mouse death must be confirmed by a combination of criteria including, lack of breathing and heart rate. Follow local animal care guidelines for euthanasia of animals.
4. Modifications have been described about the CS preparation and long-term storage. CS may be harvested from different donors, pooled, suspended in 5% dextrose, filtered and stored at −80 °C in 15% glycerol. Frozen CS should be rapidly thawed and used immediately for intraperitoneal administration to induce sepsis [22, 48].
5. Maternal neglect and cannibalism are additional challenges, aside from the technical difficulties associated with using neonatal mice. Some rodent mothers will reject their pups after being manipulated (smearing the pups with bedding can be helpful before reintroducing the mother). After returning injected pups to the cage with the mother, observe closely the mother's behavior for the first 10–15 min to confirm that it has accepted them back. The milk spot (presence of milk in the neonatal mice stomachs) is an excellent gauge of a healthy pup; therefore, when little or no milk is present then something is wrong with it or the mother has neglected it. The milk spot is clearly visible through their translucent bodies.
6. Reapply every 24 h.
7. The required lethal dose (LD) of the CS may be standardized by each laboratory to achieve the desired experimental outcomes. Significant variations occur in the intestinal microbiota content of CS donor mice potentially impacting the challenge dose necessary to induce similar LDs from one facility to another.
8. Neonatal mice that experience CS leakage or bleeding may be excluded from the study. Excluding these animals helps to ensure a uniform CS dosage.
9. In general, post administration analgesics are not recommended after induction of CS sepsis. Although these studies are usually classified as USDA Category E (unrelieved pain and distress), use of analgesics are often not helpful and can be detrimental. Opioids in general will depress spontaneous activity and especially feeding. In most cases, local animal care and use committees will permit withholding analgesics as long as there are preset criteria for determining an irreversible moribund state. Physiological or behavioral signs will define humane endpoints. Animals experiencing pain or distress

should be euthanized promptly to minimize discomfort (humane endpoint).

Acknowledgments

Supported in part by grants R01 GM097531 (SDL/LLM), P50 GM111152 (LLM), R01 GM113945 (PAE) awarded by the National Institute of General Medical Sciences, USPHS. JCR was partially supported by the Robert H. and Kathleen M. Axline Fund.

References

1. Cohen J, Vincent JL, Adhikari NK et al (2015) Sepsis: a roadmap for future research. Lancet Infect Dis 15:581–614
2. Shane AL, Sanchez PJ, Stoll BJ (2017) Neonatal sepsis. Lancet 390:1770–1780
3. Fleischmann C, Scherag A, Adhikari NK et al (2016) Assessment of global incidence and mortality of hospital-treated Sepsis. Current estimates and limitations. Am J Respir Crit Care Med 193:259–272
4. Wynn JL (2016) Defining neonatal sepsis. Curr Opin Pediatr 28:135–140
5. Raymond SL, Stortz JA, Mira JC et al (2017) Immunological defects in neonatal sepsis and potential therapeutic approaches. Front Pediatr 5:14
6. Fleischmann-Struzek C, Goldfarb DM, Schlattmann P et al (2018) The global burden of paediatric and neonatal sepsis: a systematic review. Lancet Respir Med 6:223–230
7. Paoli CJ, Reynolds MA, Sinha M et al (2018) Epidemiology and costs of sepsis in the United States-an analysis based on timing of diagnosis and severity level. Crit Care Med 46:1889–1897
8. Fink MP (2014) Animal models of sepsis. Virulence 5:143–153
9. Seok J, Warren HS, Cuenca AG et al (2013) Genomic responses in mouse models poorly mimic human inflammatory diseases. Proc Natl Acad Sci U S A 110:3507–3512
10. Efron PA, Mohr AM, Moore FA et al (2015) The future of murine sepsis and trauma research models. J Leukoc Biol 98:945–952
11. Gentile LF, Nacionales DC, Lopez MC et al (2014) Host responses to sepsis vary in different low-lethality murine models. PLoS One 9:e94404
12. Marshall JC, Deitch E, Moldawer LL et al (2005) Preclinical models of shock and sepsis: what can they tell us? Shock 24(Suppl 1):1–6
13. Stortz JA, Raymond SL, Mira JC et al (2017) Murine models of sepsis and trauma: can we bridge the gap? ILAR J 58:90–105
14. Deitch EA (1998) Animal models of sepsis and shock: a review and lessons learned. Shock 9:1–11
15. Lewis AJ, Seymour CW, Rosengart MR (2016) Current murine models of sepsis. Surg Infect 17:385–393
16. Wichterman KA, Baue AE, Chaudry IH (1980) Sepsis and septic shock—a review of laboratory models and a proposal. J Surg Res 29:189–201
17. Osuchowski MF, Ayala A, Bahrami S et al (2018) Minimum quality threshold in pre-clinical sepsis studies (MQTiPSS): an international expert consensus initiative for improvement of animal modeling in Sepsis. Shock 50:377–380
18. Nemzek JA, Hugunin KM, Opp MR (2008) Modeling sepsis in the laboratory: merging sound science with animal well-being. Comp Med 58:120–128
19. Traeger T, Koerner P, Kessler W et al (2010) Colon ascendens stent peritonitis (CASP)—a standardized model for polymicrobial abdominal sepsis. J Vis Exp 46:e2299
20. Wynn JL, Scumpia PO, Delano MJ et al (2007) Increased mortality and altered immunity in neonatal sepsis produced by generalized peritonitis. Shock 28:675–683

21. Gentile LF, Nacionales DC, Lopez MC et al (2014) Protective immunity and defects in the neonatal and elderly immune response to sepsis. J Immunol 192:3156–3165
22. Starr ME, Steele AM, Saito M et al (2014) A new cecal slurry preparation protocol with improved long-term reproducibility for animal models of sepsis. PLoS One 9:e115705
23. Rincon JC, Cuenca AL, Raymond SL et al (2018) Adjuvant pretreatment with alum protects neonatal mice in sepsis through myeloid cell activation. Clin Exp Immunol 191:268–278
24. Wynn JL, Scumpia PO, Winfield RD et al (2008) Defective innate immunity predisposes murine neonates to poor sepsis outcome but is reversed by TLR agonists. Blood 112:1750–1758
25. Cuenca AG, Wynn JL, Kelly-Scumpia KM et al (2011) Critical role for CXC ligand 10/CXC receptor 3 signaling in the murine neonatal response to sepsis. Infect Immun 79:2746–2754
26. Cuenca AG, Cuenca AL, Gentile LF et al (2015) Delayed emergency myelopoiesis following polymicrobial sepsis in neonates. Innate Immun 21:386–391
27. Fallon EA, Chun TT, Young WA et al (2017) Program cell death receptor-1-mediated invariant natural killer T-cell control of peritoneal macrophage modulates survival in neonatal Sepsis. Front Immunol 8:1469
28. Young WA, Fallon EA, Heffernan DS et al (2017) Improved survival after induction of sepsis by cecal slurry in PD-1 knockout murine neonates. Surgery 161:1387–1393
29. Remick DG, Newcomb DE, Bolgos GL et al (2000) Comparison of the mortality and inflammatory response of two models of sepsis: lipopolysaccharide vs. cecal ligation and puncture. Shock 13:110–116
30. Mathiak G, Szewczyk D, Abdullah F et al (2000) An improved clinically relevant sepsis model in the conscious rat. Crit Care Med 28:1947–1952
31. Zantl N, Uebe A, Neumann B et al (1998) Essential role of gamma interferon in survival of colon ascendens stent peritonitis, a novel murine model of abdominal sepsis. Infect Immun 66:2300–2309
32. Maier S, Traeger T, Entleutner M et al (2004) Cecal ligation and puncture versus colon ascendens stent peritonitis: two distinct animal models for polymicrobial sepsis. Shock 21:505–511
33. Kinasewitz GT, Chang AC, Peer GT et al (2000) Peritonitis in the baboon: a primate model which stimulates human sepsis. Shock 13:100–109
34. Kazarian KK, Perdue PW, Lynch W et al (1994) Porcine peritoneal sepsis: modeling for clinical relevance. Shock 1:201–212
35. DeSantis DT, Schmaltz LW (1984) The mother-litter relationship in developmental rat studies: cannibalism vs caring. Dev Psychobiol 17:255–262
36. Remick DG, Ward PA (2005) Evaluation of endotoxin models for the study of sepsis. Shock 24(Suppl 1):7–11
37. Chen P, Stanojcic M, Jeschke MG (2014) Differences between murine and human sepsis. Surg Clin North Am 94:1135–1149
38. Deitch EA (2005) Rodent models of intra-abdominal infection. Shock 24(Suppl 1):19–23
39. Turnbull IR, Wlzorek JJ, Osborne D et al (2003) Effects of age on mortality and antibiotic efficacy in cecal ligation and puncture. Shock 19:310–313
40. Brown I, Bellevue O, Shawo A et al (2015) Low-dose cyclophosphamide improves survival in a murine treatment model of sepsis. Shock 43:92–98
41. Steele AM, Starr ME, Saito H (2017) Late therapeutic intervention with antibiotics and fluid resuscitation allows for a prolonged disease course with high survival in a severe murine model of sepsis. Shock 47:726–734
42. Denning NL, Yang WL, Hansen L et al (2019) C23, an oligopeptide derived from cold-inducible RNA-binding protein, suppresses inflammation and reduces lung injury in neonatal sepsis. J Pediatr Surg 54(10):2053–2060
43. Ganatra HA, Varisco BM, Harmon K et al (2017) Zinc supplementation leads to immune modulation and improved survival in a juvenile model of murine sepsis. Innate Immun 23:67–76
44. Sellers RS, Clifford CB, Treuting PM et al (2012) Immunological variation between inbred laboratory mouse strains: points to consider in phenotyping genetically immunomodified mice. Vet Pathol 49:32–43
45. Gentile LF, Cuenca AL, Cuenca AG et al (2015) Improved emergency myelopoiesis

and survival in neonatal sepsis by caspase-1/11 ablation. Immunology 145:300–311

46. Wynn JL, Wilson CS, Hawiger J et al (2016) Targeting IL-17A attenuates neonatal sepsis mortality induced by IL-18. Proc Natl Acad Sci U S A 113:E2627–E2635
47. Cuenca AG, Joiner DN, Gentile LF et al (2015) TRIF-dependent innate immune activation is critical for survival to neonatal gram-negative sepsis. J Immunol 194:1169–1177
48. Brook B, Amenyogbe N, Ben-Othman R et al (2019) A controlled mouse model for neonatal polymicrobial sepsis. J Vis Exp (143)

Chapter 5

Injection of *Escherichia coli* to Induce Sepsis

Xian-Hui He, Dong-Yun Ouyang, and Li-Hui Xu

Abstract

Mouse models of bacterial sepsis are widely used in research to investigate the underlying molecular mechanisms of sepsis and to develop clinically useful therapeutic regimens. Three commonly used mouse sepsis models include (a) injection of bacterial endotoxin, (b) infusion of cultured bacteria, and (c) cecal ligation and puncture. Here we describe the induction of bacterial sepsis in mice by intraperitoneal injection of cultured live *Escherichia coli* cells. The severity of the sepsis can be regulated by the number of *E. coli* cells injected into the peritoneal cavity of mice.

Key words Bacterial sepsis, *Escherichia coli*, Intraperitoneal injection, Colony-forming units, Survival curve, Interleukin-1β, Tissue damage

1 Introduction

Sepsis was traditionally identified as a severe inflammatory response to overwhelming infection, defined as a systemic inflammatory response syndrome (SIRS) [1]. The definition of sepsis has recently been revised to identify patients with life-threatening organ dysfunction caused by a dysregulated host response to infection [2]. Sepsis is the leading cause of death in critically ill patients. We currently have no effective therapy for sepsis, except supportive treatments and antibiotic regimens [1]. It is therefore extremely valuable to use animal models of sepsis for the development of new therapeutic strategies. Three mouse models of sepsis have been commonly used in preclinical studies: (a) injection of bacterial endotoxin (e.g., lipopolysaccharide [LPS]); (b) infusion or instillation of cultured bacteria; and (c) disruption of the endogenous protective barrier, for example by cecal ligation and puncture (CLP) [3]. Although LPS-induced endotoxemic shock is widely used in research, the latter two approaches may be more relevant to clinical patients with infectious sepsis. Notably, CLP is still the most widely used animal model of polymicrobial sepsis. However, the surgical manipulation is not well standardized, and disease severity

Wendy E. Walker (ed.), *Sepsis: Methods and Protocols*, Methods in Molecular Biology, vol. 2321,
https://doi.org/10.1007/978-1-0716-1488-4_5,

may vary due to the length of cecal ligation and the size of the needle puncture, among other factors [4]. On the other hand, intraperitoneal injection of viable bacteria from culture constitutes an alternative mouse model of sepsis with high reproducibility, and has therefore been adopted by many researchers as a model of bacterial sepsis [5, 6].

Based on several previous studies, we have established a bacterial sepsis model in mice for investigating therapeutic agents from herbal medicines [7, 8]. Live *Escherichia coli* cells are freshly prepared and 1–2.5 $\times$ 10^9 colony-forming units (CFU) of live *E. coli* bacteria are injected into the peritoneal cavity of each mouse. Mouse survival is monitored every 6 h for 4–5 consecutive days (Fig. 1a). A systemic inflammatory response is observed in mice with bacterial sepsis, evidenced by elevated levels of interleukin (IL)-1β in the blood and in the peritoneal cavity (Figs. 1b and 2c). Tissue damage is observed in the liver, accompanied by infiltration of inflammatory cells (Figs. 1c and 2b). The severity of sepsis can be adjusted by changing the bacterial CFU of live *E. coli* bacteria administered to each mouse (Fig. 2a) [9].

2 Materials

All media and solutions are prepared with deionized water and analytical grade reagents. The media and solutions are autoclaved immediately and stored at room temperature. Follow all waste disposal regulations when disposing waste materials. Animal experiments are designed following the National Institutes of Health guidelines and were approved by the Committee on the Ethics of Animal Experiments of Jinan University (Guangzhou).

2.1 Equipment

1. Rotary shaker incubator.
2. Stationary incubator.
3. Centrifuge with rotor for 50 mL conical tubes, capable of an 1824 $\times$ *g*-force.
4. Spectrophotometer, capable of reading 10 mm cuvettes at 600 nm.

2.2 Sterile Plastic and Glassware

1. Disposable inoculation loops.
2. 90-mm plastic plates.
3. 50 mL plastic tubes.
4. 200 mL conical flasks covered with tinfoil, for bacterial culture.
5. 1 mL syringes.
6. 26-G needles.
7. Cotton swabs.

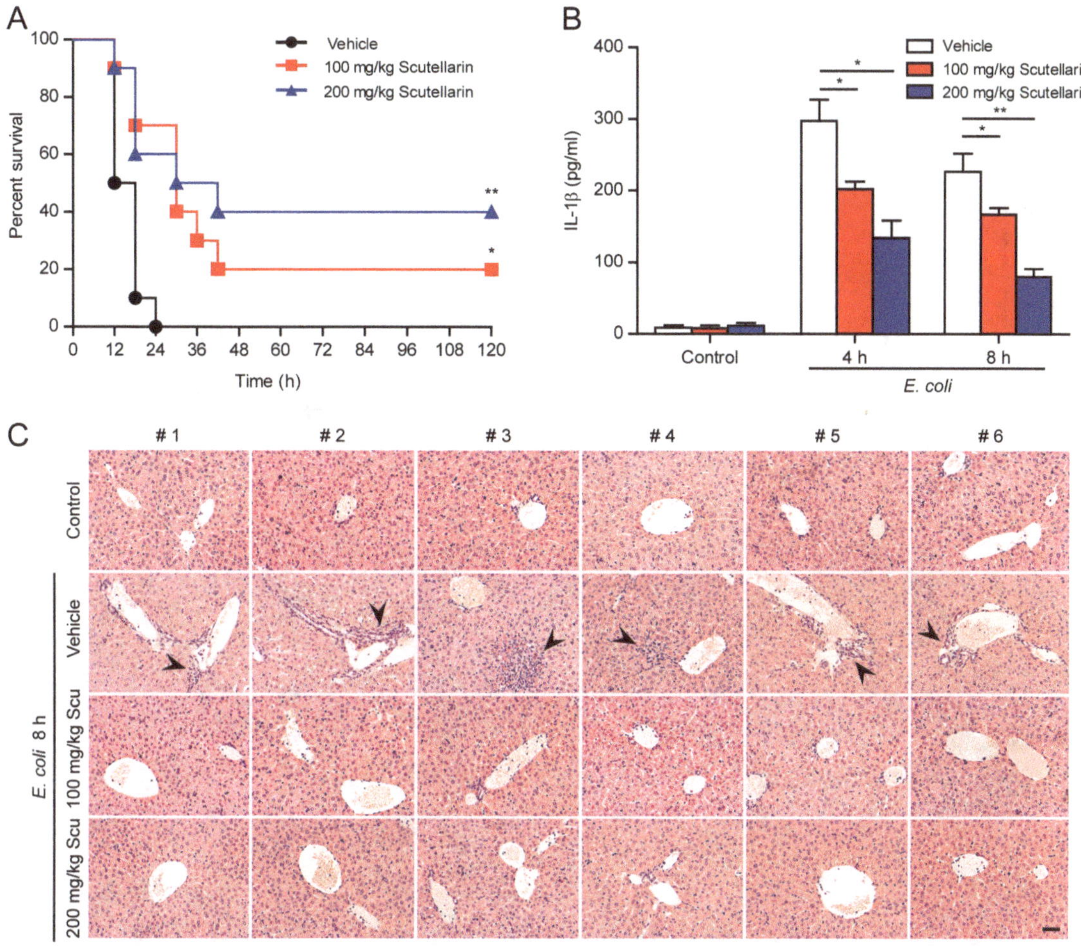

Fig. 1 Induction of bacterial sepsis by intraperitoneal injection of Escherichia coli. C57BL/6 mice were administered scutellarin (100 or 200 mg/kg body weight) or vehicle (2% Tween 80 in PBS) once intragastrically (i.g.) 3 h prior to bacterial infection intraperitoneally (i.p.) with viable E. coli (2.5×10^9 CFU/mouse). Mice were administered scutellarin or vehicle i.g. once again at 1 h after the bacterial injection. (**a**) Mouse survival was monitored every 6 h for 5 consecutive days. Kaplan–Meier survival curves were used to analyze the data (10 mice per group). The significance was evaluated by the log-rank (Mantel–Cox) test. **(b)** Mice were treated as in (**a**) except that mice were injected with viable E. coli (2.0×10^9 CFU/mouse i.p.). The serum levels of IL-1β at 4 h and 8 h post–bacterial infection were measured by bead-based assay (6 mice per group). Data are shown as mean ± SD ($n = 6$). $^*P < 0.05$; $^{**}P < 0.01$. (**c**) Hematoxylin and eosin staining of mouse livers harvested 8 h after infection with live E. coli. Representative images from each mouse are presented and arrowheads indicate infiltrating inflammatory cells. The number on the top indicates the mouse number. Scale bar, 50 μm; Scu, scutellarin. (reproduced from ref. 8 with the authors' owned copyright)

2.3 Reagents

1. *E. coli* strain DH5α frozen stock (*see* **Note 1**).
2. Luria–Bertani (LB) liquid medium: 10 g tryptone, 5 g yeast extract, 10 g NaCl, and 950 mL deionized H_2O. Stir the solution until the solutes have dissolved. Add deionized H_2O till the volume of the solution is 1 L. Autoclave the solution for 20 min at 15 psi ($1.05\ kg/cm^2$) to sterilize it.

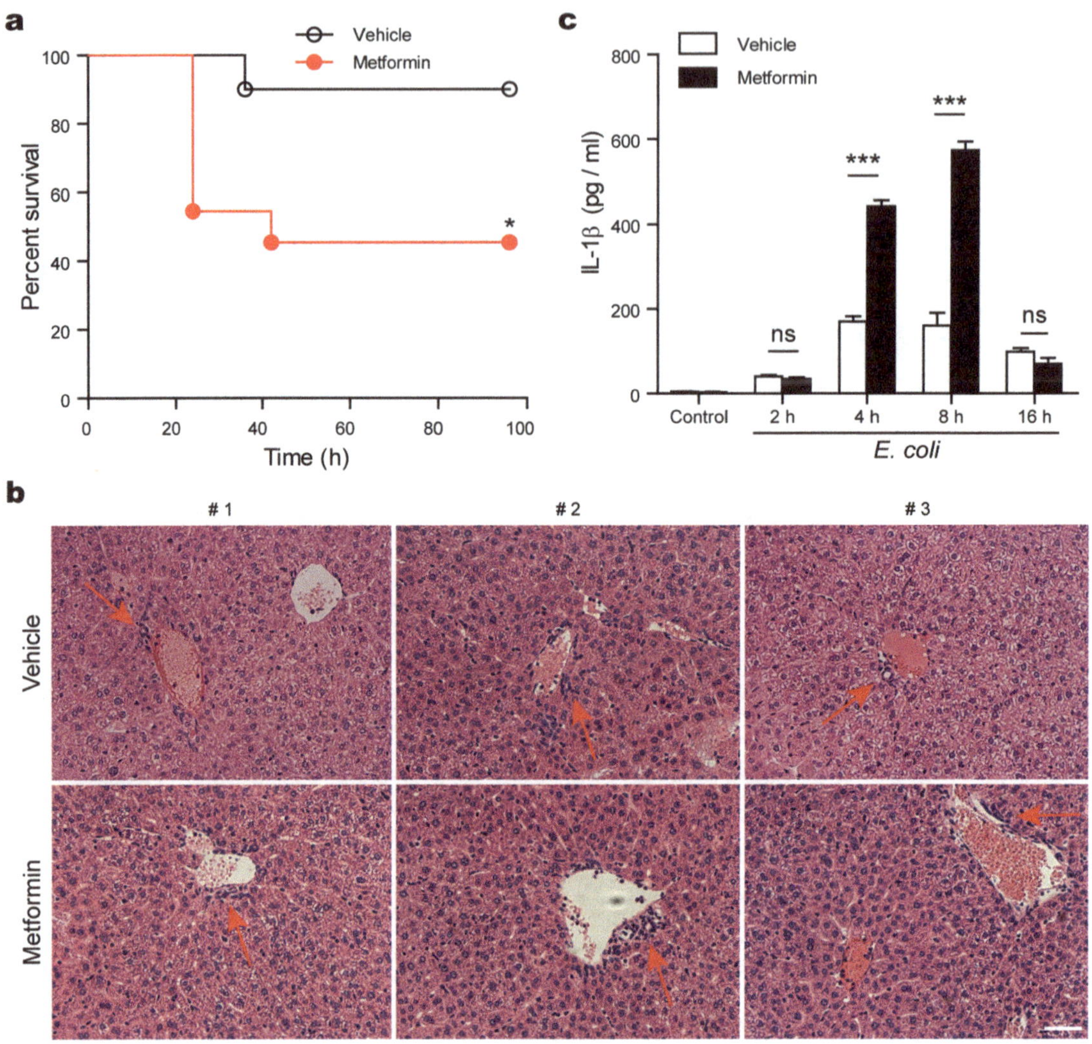

Fig. 2 Metformin administration aggravated bacterial sepsis in mice. C57BL/6 mice were injected with viable E. coli at a dose of 1 × 10^9 CFU/mouse i.p. After 1 h, the mice were administered metformin at a dose of 250 mg/kg body weight or vehicle i.g. (**a**) Kaplan-Meier survival curves. Each group contained 10 mice and three independent experiments were performed, with one set of representative data presented. $^{*}P < 0.05$. (**b**) Histopathological analysis of liver sections after bacterial infection. Red arrows indicate infiltrating inflammatory cells. (**c**) Serum IL-1β levels after bacterial infection. Data are shown as mean ± SD ($n = 5$). Each group contained five mice and two independent experiments were performed, with one set of data presented. Scale bar, 50 μm. $^{***}P < 0.001$; ns, not significant (reproduced from ref. 9 with the authors' owned copyright)

3. LB agar plates: 15 g Bacto agar plus LB liquid medium (described above) to make a final volume of 1 L. Autoclave for 20 min at 15 psi (1.05 kg/cm^2). After removing the medium from the autoclave, swirl it gently to disperse the melted agar evenly throughout the solution (*see* **Note 2**). Allow the medium to cool to 50–60 °C. Pour 30–35 mL of

medium per 90-mm plastic plate, using aseptic technique. After the medium has hardened completely, store the plates inverted at 4 °C until use (*see* **Note 3**).

4. Phosphate-buffered saline (PBS): 3.484 g of $Na_2HPO_4.12H_2O$, 8 g of NaCl, 0.2 g of KCl, and 0.2 g of KH_2PO_4 and 900 mL of deionized H_2O. Stir to dissolve. Add each salt in series after the previous one has completely dissolved. Adjust pH to 7.4. Add deionized H_2O to make a final volume of 1 L. Autoclave PBS for 20 min at 15 psi (1.05 kg/cm^2) to sterilize it.
5. 70% ethanol.

2.4 Experimental Animals

1. C57BL/6 mice: 6–8 weeks of age with a body weight of approximately 18 g. Acclimatize animals for 1 week before completing the experiments.

2.5 Histopathology and Serum Cytokine Analysis

1. 10% formalin solution: to fix organs for histopathological analysis.
2. Small glass capillary tubes: to collect blood for serum preparation.

3 Methods

3.1 Preparation of Live Bacteria

All bacteria preparation procedures should be performed using good aseptic technique and sterile, disposable tools and plasticware.

1. Using a sterile disposable inoculation loop, obtain a small amount of the frozen *E. coli* DH5α bacteria stock solution (*see* **Note 1**) and streak it onto an LB agar plate. Culture the plate at 37 °C for 24 h in a stationary incubator to allow separate colonies to form.
2. Obtain a single colony from the LB agar plate and inoculate into a 50 mL tube containing 3 mL of LB liquid medium. Culture the tube in a rotatory shaker incubator (200 RPM) at 37 °C overnight.
3. Inoculate 1 mL of the overnight bacterial culture into a 200 mL conical bottle with 50 mL of LB liquid medium. Culture the bottle in a rotary shaker incubator (200 RPM) at 37 °C for 3–4 h.
4. Transfer the culture into a 50 mL conical tube and pellet the bacteria by centrifugation at 1824 × *g* for 10 min. Wash the bacteria twice with 30 mL of PBS. For each wash, resuspend the pellet in 30 mL PBS by pipetting and vortex mixing, centrifuge as described above, and then dispose of the supernatant. After the second wash, resuspended the pellet in 20 mL of PBS by pipetting and vortex mixing.

3.2 CFU Determination

1. Determine the concentration of bacteria in the suspension with a spectrophotometer, by measuring the optical density (OD) at 600 nm with a light path of 10 mm, using PBS as a blank control (*see* **Note 4**).
2. The concentration of CFU-forming bacteria in the solution can be determined more accurately, in a retrospective fashion (*see* **Note 5**). To determine CFU, make serial dilutions of the *E. coli* suspension in 1.5 mL tubes with LB liquid medium. The final three dilutions should have an expected concentration of 1000–3000 cells/mL, based on the value from spectrophotometry.
3. Aliquot 100 μL of liquid from each dilution tube onto a separate LB plate, and spread the liquid evenly over the plate. When the plate has completely dried, incubate it inverted at 37 °C overnight. Each live bacterium will grow into a separate colony on the plate, usually within 24 h.
4. Count the total colonies on each plate with a grid paper under the plate. The CFU/mL of the *E. coli* suspension can be calculated by multiplying the number of colonies on the plate by 10× the dilution factor used for the plate.

3.3 Intraperitoneal Injection of E. coli

1. Adjust the *E. coli* solution to the desired concentration. A 5×10^9 CFU/mL solution will induce rapid mortality in all animals, while a $2–3 \times 10^9$ CFU/mL solution will induce a lower mortality rate (*see* **Note 6**). Use a freshly prepared solution for each experiment (*see* **Note 5**).
2. For each mouse, prepare a sterile 1 mL syringe with a 26-G needle.
3. Load 0.5 mL of fully resuspended *E. coli* solution ($2–5 \times 10^9$ CFU/mL) into the syringe. Remove any air bubbles.
4. Remove the mouse from the cage using gentle handling and manually restrain the mouse (*see* **Note 7**).
5. Swab the abdominal injection site with 70% ethanol on a cotton swab.
6. Inject 0.5 mL of *E. coli* suspension into the peritoneal cavity of the mouse (*see* **Note 8**).
7. Monitor mouse survival every 6 h for 4–5 consecutive days. Mouse survival should analyzed using Kaplan-Meier survival curves. The statistical difference between 2 groups can be determined using the log-rank (Mantel-Cox) test.

3.4 Serum Cytokine Analysis

1. Collect blood from the retro-orbital plexus (*see* **Note 9**).
2. Prepare serum by coagulation and centrifugation (*see* **Notes 10**).
3. Measure cytokines in the serum via commercially available ELISA kits or bead assays (*see* **Note 11**).

3.5 Histopathology

1. Euthanize mice (*see* **Notes 9** and **12**).
2. After death has been confirmed, harvest the liver and colon tissues.
3. Fix the liver and colon tissues in 10% neutral formalin for 24 h.
4. Use a commercial service to obtain paraffin slices of the liver and colon, and stain them with hematoxylin and eosin (*see* **Note 13**).
5. Visualize the tissues under a microscope armed with a digital color camera.

4 Notes

1. This protocol describes a standard procedure to induce sepsis in mice using the *E. coli* DH5α strain. Different strains of *E. coli* (e.g., strains JM109 and 0111:B4) [10, 11] and isolated *Enterococcus faecalis* [5, 6] can also be used to induce sepsis in mice.
2. Be careful: the medium may boil over when swirled, as the solution is very hot. Avoid producing air bubbles during swirling.
3. Remove the plates from storage 1–2 h before they are used. To avoid "sweat", the plates can be incubated at 37 °C for several hours in an inverted position before they are used.
4. The value of absorbance (A) at 600 nm will depend on the light path of the cuvette and will vary among spectrophotometers. If the light path of a spectrophotometer is 1 mm, a factor of 10 is used to calculate the OD_{600} value. As a good rule of thumb, each 0.1 OD unit is roughly equivalent to 1×10^8 cells/mL. Calculate the number of cells/mL from a suspension that has an $OD_{600} < 1$. The bacteria from 50 mL of LB liquid medium may have ~4–6 × 10^{10} CFU in total, depending on the inoculation volume of the overnight culture.
5. As freshly prepared bacteria are immediately used to induce sepsis, we usually use spectrophotometry for timely determination of the CFU of *E. coli* based on the empirical formula from CFU determination in combination with spectrophotometry: $1.0\ OD_{600} = 1 \times 10^9$ CFU/mL.
6. We routinely inject 0.5 mL of bacterial suspension in PBS into the peritoneal cavity of each mouse. Ensure the bacteria are evenly suspended by gently vortexing. Injection of 2–2.5 × 10^9 CFU of live *E. coli* cells into the peritoneal cavity is a lethal dose resulting in 100% mortality within 24–36 h. Injection of 1–1.5 × 10^9 CFU of live *E. coli* cells into the peritoneal cavity is a sub-lethal dose culminating in <50% mortality within

5 days. The sub-lethal dose of bacteria is adopted to analyze histopathology of the liver and the colon, and to determine serum cytokine levels beyond 24 h. Agents that exacerbate bacterial sepsis may also be evaluated in mice injected with the sub-lethal dose of live bacteria.

7. The handling and restraint of mice should be done gently, avoiding loud noises which are stressful to the mice. First, remove a mouse from the cage by lifting its tail with your dominant hand. Place it on a wire-bar cage lid to allow the mouse to grasp onto the bars. Second, approach the neck of the mouse from the rear with your free, nondominant hand and tightly gather the skin behind the ears with the thumb and index finger. Third, transfer the tail from your dominant hand to a position beneath the little finger of the nondominant hand, while continuing to grasp the scruff of the neck. Inject the restrained mouse via the intraperitoneal route.
8. Insert the needle into the lower left or right quadrant of the abdomen, avoiding the abdominal midline. Inject into the peritoneal cavity with moderate pressure and speed. Avoid injecting subcutaneously or inserting the needle too deeply, which could pierce organs in the peritoneal cavity.
9. Time points of 2, 4, 8, 16, and 24 h after *E. coli* injection are commonly used to assess cytokine levels in the blood or in the peritoneal cavity. A time point of 8 h after *E. coli* injection is commonly used to examine tissue damage and infiltration of inflammatory cells.
10. To prepare serum, allow the blood samples to coagulate by incubation at room temperature for 1 h, followed by incubation at 4 °C overnight. Centrifuge the samples at 900 $\times$ *g* at 4 °C for 30 min and collect the liquid (serum) fraction.
11. Cytokine levels can be determined by commercially available ELISA kits, measured with a microplate reader or a bead-based assay kit, measured through flow cytometry.
12. Use a CO_2 chamber to euthanize mice. Place the animal inside the chamber and fill with the CO_2 at 1 L/min. Wait approximately 3–5 min for the animal to stop moving or breathing. Eyes should be fixed and dilated.
13. Both paraffin sectioning and hematoxylin and eosin staining are multistep procedures requiring high expertise. We prefer to use a commercial service for this histopathological analysis.

Acknowledgments

This work was supported by the grants from the National Natural Science Foundation of China (Nos. 81773965, 81673664, and 81373423).

References

1. Angus DC, van der Poll T (2013) Severe sepsis and septic shock. N Engl J Med 369:2063
2. Singer M, Deutschman CS, Seymour CW et al (2016) The third international consensus definitions for sepsis and septic shock (sepsis-3). JAMA 315:801–810
3. Doi K, Leelahavanichkul A, Yuen PS et al (2009) Animal models of sepsis and sepsis-induced kidney injury. J Clin Invest 119:2868–2878
4. Dejager L, Pinheiro I, Dejonckheere E et al (2011) Cecal ligation and puncture: the gold standard model for polymicrobial sepsis? Trends Microbiol 19:198–208
5. Wegiel B, Larsen R, Gallo D et al (2014) Macrophages sense and kill bacteria through carbon monoxide-dependent inflammasome activation. J Clin Invest 124:4926–4940
6. Chung SW, Liu X, Macias AA et al (2008) Heme oxygenase-1-derived carbon monoxide enhances the host defense response to microbial sepsis in mice. J Clin Invest 118:239–247
7. Pan H, Xu LH, Huang MY et al (2015) Piperine metabolically regulates peritoneal resident macrophages to potentiate their functions against bacterial infection. Oncotarget 6:32468–32483
8. Liu Y, Jing YY, Zeng CY et al (2017) Scutellarin suppresses NLRP3 inflammasome activation in macrophages and protects mice against bacterial Sepsis. Front Pharmacol 8:975
9. Zha QB, Wei HX, Li CG et al (2016) ATP-induced inflammasome activation and pyroptosis is regulated by AMP-activated protein kinase in macrophages. Front Immunol 7:597
10. Ren Y, Hua L, Meng X et al (2016) Correlation of surface toll-like receptor 9 expression with IL-17 production in neutrophils during septic peritonitis in mice induced by E. coli. Mediators Inflamm 2016:3296307
11. Long C, Wang Y, Herrera AH et al (2010) In vivo role of leukocyte ADAM17 in the inflammatory and host responses during E. coli-mediated peritonitis. J Leukoc Biol 87:1097–1101

Chapter 6

A Mouse Model of *Pseudomonas aeruginosa* Pneumonia

Brian W. LeBlanc and Craig T. Lefort

Abstract

The rapid innate immune response to respiratory infection is essential to prevent the systemic dissemination of pathogens. This chapter outlines an experimental mouse model of respiratory infection by gram-negative *Pseudomonas aeruginosa* and analyses of leukocyte trafficking in the lungs. The reader will learn two methods to induce respiratory infection in mice that differ in whether the initial bolus is targeted within a specific lobe of the lung. We then describe a technique based on tissue digestion and flow cytometry that allows the investigator to distinguish leukocytes within different compartments of the lung, and discuss the advantages and limitations to such an approach.

Key words Inflammation, Acute lung injury, Bacterial infection, Pneumonia, Neutrophils, Granulocytes, Sepsis, Fluorescence-activated cell sorting (FACS), Intratracheal

1 Introduction

Nosocomial infections of the respiratory system occur frequently in the intensive care unit (ICU), as immune suppression and dysfunction are key features of the response to trauma, hemorrhage, and/or major surgery [1]. The mortality rate attributed to nosocomial pneumonia in the ICU is as high as 50% [2]. *Pseudomonas aeruginosa* is one of the most common bacterial species observed in ICU-acquired and/or ventilator-associated pneumonia [3]. By contrast, *P. aeruginosa* is rarely observed causing community-acquired pneumonia [4], indicating that underlying host defense dysfunction associated with critical illness is a prominent factor in susceptibility to lower respiratory infection by *P. aeruginosa*. As such, investigation of the pulmonary inflammatory response to *P. aeruginosa* using experimental animal models is critical for the development of therapeutic strategies to address this clinical challenge.

In this chapter, we describe methods to instill *P. aeruginosa* into the lungs of mice, including a variation of delivery that allows for additional mechanistic insight into the host response. The latter

Wendy E. Walker (ed.), *Sepsis: Methods and Protocols*, Methods in Molecular Biology, vol. 2321,
https://doi.org/10.1007/978-1-0716-1488-4_6, © Springer Science+Business Media, LLC, part of Springer Nature 2021

technique involves the targeting of intratracheal *P. aeruginosa* to the left lung exclusively. Reasons why an investigator may want to choose this targeted approach include the following: (1) The right lobes of the lung can serve as comparison to evaluate direct versus indirect responses of host-pathogen interaction. and (2) In most cases it would be expected that gas exchange in the contralateral lung to remains intact, thus extending the experimental window.

Phagocytic leukocytes of the innate immune system, including neutrophils and macrophages, are critical to the eradication of *P. aeruginosa* from the lungs [5–7]. We have recently found that prior hemorrhagic shock predisposes mice to an exacerbated inflammatory response to subsequent *P. aeruginosa* respiratory infection in terms of cytokine generation, but one that ultimately compromises neutrophil entry into the bronchoalveolar airspaces of the lung where the infection was initiated [8]. Those findings, which employed the mouse model of *P. aeruginosa* pneumonia that is described here, are consistent with the concept that it is the neutrophils that become sequestered in the extensive pulmonary capillary bed that mediate acute lung injury, rather than those neutrophils that enter the alveoli [9]. These studies underscore the importance of probing the tissue/airspace versus intravascular localization of immune cells, as will be described in this chapter. This in vivo intravascular labeling approach has been described generally and more extensively elsewhere [10], and a specific application to analyzing neutrophils in the infected lung will be presented here.

2 Materials

2.1 Intratracheal Inhalation

1. *Pseudomonas aeruginosa* (*see* **Note 1**).
2. Tryptic soy broth (TSB), sterile (autoclaved).
3. TSB/15% agar plates.
4. Sterile inoculating loop or pipette tip.
5. 0.9% saline, USP.
6. Spectrophotometer.
7. Centrifuge.
8. Isoflurane.
9. Isoflurane vaporizer apparatus.
10. Rodent intubation stand.
11. Graefe curved forceps.

2.2 Targeted Intratracheal Administration

1. Isoflurane.
2. Isoflurane vaporizer apparatus.

3. Mouse anesthesia nose cone.
4. Fine dissection scissors, sharp/sharp.
5. Graefe curved forceps.
6. Silk suture, 3–0.
7. 24-gauge Angiocath.

2.3 Lung Processing and Flow Cytometry

1. Mouse Lung Dissociation Kit (Miltenyi, catalog #130-095-927).
2. Phycoerythrin (PE)-conjugated rat anti-CD45 antibody.
3. Isoflurane.
4. Isoflurane vaporizer apparatus.
5. Insulin syringe.
6. (2) Graefe curved forceps.
7. Dissection scissors.
8. Phosphate-buffered saline (PBS), pH 7.4.
9. PEB buffer: PBS, 2 mM EDTA, 0.5% bovine serum albumin (BSA).
10. Allophycocyanin (APC)-conjugated rat anti-Ly6G antibody.
11. Flow cytometer.

3 Methods

3.1 Intratracheal Inhalation

1. Using standard microbiological techniques (*see* **Note 2**), spread an inoculum of *P. aeruginosa* from a frozen stock on a TSB/agar plate and grow 16–24 h at 37 °C.
2. Pick a single colony of *P. aeruginosa* from the TSB/agar plate and inoculate into 3 mL of sterile tryptic soy broth in a 14 mL Falcon tube. Grow 16–20 h in a heated shaker at 250 RPM, 37 °C.
3. Transfer 1 mL of dense *P. aeruginosa* liquid culture to a 1.5-mL microcentrifuge tube. Wash the sample by centrifugation at $1000 \times g$ (rotor radius 8.4 cm, 3400 RPM) for 5 min, aspirating the supernatant, and then resuspending the bacterial pellet in 1 mL 0.9% saline. Repeat for a total of 3 washes.
4. To determine the number of colony forming units (CFUs) in the sample using a spectrophotometer, measure the sample's optical density at 540 nm (OD_{540}). If the OD_{540} is outside of the linear range (*see* **Note 3**), dilute the sample accordingly and repeat the OD_{540} measurement.
5. Make a stock solution of 0.9% saline with the desired concentration of *P. aeruginosa* to achieve delivery of the desired dose per 50 μL volume. For strain PA103, we have empirically

determined that 3×10^5 CFU is predominantly sublethal in control mice, while it induces a mortality rate of ~60% in mice receiving prior hemorrhagic shock [8].

6. Anesthetize an 8–12-week-old laboratory mouse (*see* **Note 4**) using 2–4% isoflurane in oxygen gas.
7. Once sedated, position the mouse on an intubation stand by placing the animal's front teeth over the horizontally mounted rubber band. Use a pair of forceps to gently hold the mouse's tongue in place while opening the oral cavity. Pipet a 50 μL bolus of the prepared *P. aeruginosa* stock solution into the back of the oral cavity at the base of the tongue (*see* **Note 5**). Immediately and gently pinch the animal's nose to force the inhalation of the bolus during the next inspiration while visually confirming that the bolus disappears from the rear of the oral cavity.
8. Remove the mouse from the intubation stand. Hold the mouse upright for approximately 30 s as it resumes normal breathing and recovers from anesthesia. Return the mouse to its cage. Continue to monitor the mouse (according to your approved institutional protocol) until the experimental endpoint.

3.2 Targeted Intratracheal Administration

1. Prepare *P. aeruginosa* as described above.
2. Anesthetize the mouse using 2–4% isoflurane in oxygen gas. Maintain 1–2% isoflurane anesthesia during this procedure (except where noted otherwise) using a nose cone placed over the animal's snout.
3. Place the mouse in the supine position on a sterile surgical surface (*see* **Note 6**) with its head pointing away from you.
4. Using fine scissors, make a midline incision above the trachea from ear level to the top of the sternum (*see* **Note 7**).
5. Using curved forceps, spread the parotid gland and thymus laterally away from the midline. At the midpoint of incision, press the curved forceps tip into the medial border of the sternohyoid/sternothyroid muscle creating a hole within which you then spread the muscles apart laterally to expose the tracheal sheath. Pinch the peritracheal sheath with two forceps and pull apart lengthwise to expose the underlying trachea (notice cartilage rings).
6. Position curved forceps to one side of the trachea and guide their tips underneath the trachea to other side. With your other hand, guide a 6-in. piece of silk suture toward the forceps tips. Using the forceps positioned underneath the trachea, grasp the suture and pull it through, leaving about half its length on either side of the trachea.

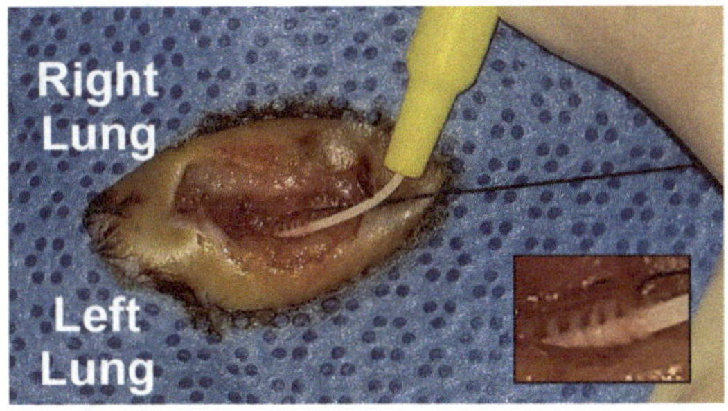

Fig. 1 Insertion of the Angiocath for targeted delivery of *P. aeruginosa*. To guide the tip of the Angiocath sheath toward the left lobe of the lung, the Angiocath base should be bent away from the left lung as it is advanced. (**Inset**) Note that the Angiocath is positioned against the left side of the trachea

7. At this point, position the mouse with its head on your dominant side (i.e., toward the right for right-handed surgeons). Align the Angiocath lengthwise above the trachea with its needle tip positioned about halfway from head to sternum and pointed toward the sternum.
8. Using its needle tip with its bevel facing up, insert the Angiocath into the trachea at a shallow angle and between two cartilage rings (*see* **Note 8**). Be careful not to lose sight of the needle tip as it is inserted just below the surface of the trachea to avoid puncturing nonvisible bronchi.
9. While holding the sheath of the Angiocath in place, remove the needle from the sheath by gently pulling the base of the needle. Advance the Angiocath sheath now within the trachea a couple millimeters so that it does not slip out of the tracheal insertion site.
10. Using the suture around the trachea as an anchor point, gently slide the Angiocath sheath into the left bronchi of the lung while bending the Luer part of the sheath base away from the left lung as it is advanced (Fig. 1). The slight bending of the sheath base will cause the sheath tip to bend toward the left bronchi by opposing pressure. Stop advancing the Angiocath when it encounters resistance. Repeat several times to ensure proper left bronchi placement (*see* **Note 9**).
11. Load a 200-μL pipettor with the prepared 50 μL bolus of *P. aeruginosa*. Press the pipette tip firmly into the Angiocath sheath base opening and quickly inject the instillate. Do not release the pipettor plunger until the tip is removed from Angiocath, so as not to withdraw the delivered bolus.
12. Quickly take the second 200-μL pipettor (with an empty tip and set to 200 μL) and inject a puff of air into the Angiocath sheath base. Again, be sure to remove the tip from the

Angiocath base each time before you release the pipettor plunger. Repeat the injection of air 2–3 times to ensure dispersal of the instillate into the left lung (*see* **Note 10**).

13. Carefully remove the Angiocath from the trachea and lift the mouse with one front paw in each hand. Raise one paw until the full weight of the mouse is pulling down, then raise the other paw similarly. Repeat 2–3 times. Ensure that the mouse resumes normal breathing.
14. Place the mouse back on the surgical surface. Use forceps to replace the muscles, glands, and skin medially covering the trachea. Use sutures to close the midline incision. Place the mouse back in its cage and monitor the mouse (according to your approved institutional protocol) until the experimental endpoint.

3.3 Lung Processing and Flow Cytometry

1. Prepare the lung dissociation buffer according to the instructions of the Miltenyi Mouse Lung Dissociation Kit. Add 2.4 mL of the lung dissociation buffer to each C-tube in which the digestion is to be performed.
2. Prepare the antibody for labeling intravascular leukocytes. Add 1 μg PE-conjugated anti-CD45 antibody (*see* **Note 11**) to 100 μL 0.9% saline and load the solution into an insulin syringe.
3. Anesthetize the mouse using 2–4% isoflurane in oxygen gas.
4. Inject the PE anti-CD45 labeling solution via retro-orbital or tail vein injection.
5. After allowing for intravascular labeling of leukocytes for 5 min, euthanize the mouse.
6. Using dissection scissors, perform a thoracotomy on the euthanized mouse to access the lungs and then use forceps to collect each lobe of the lung by pulling it apart from the trachea. Briefly rinse each lobe in PBS and then put all lobes into the C-tube containing lung dissociation buffer.
7. Follow the Miltenyi Mouse Lung Dissociation Kit instructions to digest the tissue into a single cell suspension. At the final step, the digested tissue should be resuspended in PEB buffer at a concentration of 1×10^7 cells/mL and placed on ice. For all subsequent steps, samples should be kept on ice.
8. Aliquot 100 μL of each sample into new 1.5-mL microcentrifuge tubes (or a plate or other preferred tube type); make as many new aliquots of each sample as necessary, depending on the flow cytometry panel of the investigator's design.
9. Add fluorophore-conjugated antibodies at 1–10 μg/mL concentration to samples (*see* **Note 12**). Incubate samples on ice for 30–60 min, then wash samples twice with ice-cold PEB buffer.

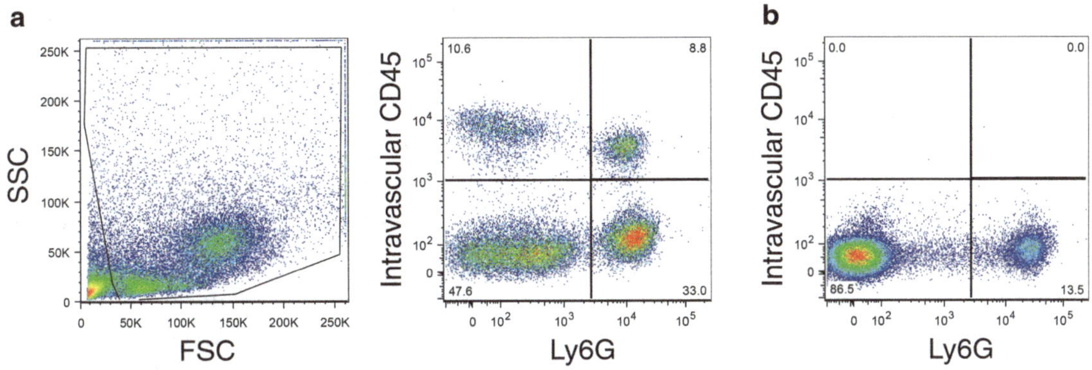

Fig. 2 FACS analysis of neutrophils in the lung of mice infected with *P. aeruginosa*. (**a**) First, use the forward scatter (FSC) and side scatter (SSC) profile to exclude red blood cells and debris from analyses of digested lung tissue. Subsequent analysis of the intravascular anti-CD45 label (y-axis) and anti-Ly6G to identify neutrophils (x-axis) reveals populations of intravascular neutrophils ($CD45^{+}Ly6G^{+}$) and extravascular neutrophils ($CD45^{-}Ly6G^{+}$). Nonneutrophil intravascular leukocytes appear in the upper left quadrant ($CD45^{+}Ly6G^{-}$). (**b**) Equivalent FACS analysis of bronchoalveolar lavage fluid indicates that no neutrophils in the airspaces were exposed to the intravascular anti-CD45 label

10. Perform flow cytometry acquisition and analysis according to standard and instrument-specific protocols. Figure 2 is an example set of flow cytometry dot-plots that represent an approach to distinguish neutrophils in the lung vasculature from those that have emigrated into the parenchyma and airspaces. In this example, intravascular leukocytes were labeled in vivo using PE anti-CD45 (as described above), and then the digested lung tissue was labeled with APC anti-Ly6G to identify neutrophils (Fig. 2a). As a control, the bronchoalveolar lavage of a *P. aeruginosa*-infected mouse receiving intravascular anti-CD45 label was collected and analyzed to demonstrate that this antibody did not leak into the airspaces (Fig. 2b).

4 Notes

1. Strains of *P. aeruginosa* are available commercially (e.g., from American Type Culture Collection). Many of these are clinical isolates that are now used as laboratory strains. A common cytotoxic strain used in many experimental models is PA103. Preservation of *P. aeruginosa* as a frozen stock is described elsewhere [11].
2. Use the appropriate biohazard precautions (BSL-2 for most *P. aeruginosa* strains) and handling procedures as outlined by institutional and/or governmental regulations and guidelines.
3. A standard curve should first be established to correlate the OD_{540} of a given strain of *P. aeruginosa* to the number of CFUs. For strain PA103, we have empirically determined that an OD_{540} of 0.25 corresponds to 2×10^8 CFU/mL.

4. All studies and procedures that involve mice should be approved by your Institutional Animal Care and Use Committee.
5. This procedure is somewhat difficult when performed by a single investigator. It is made easier if one investigator holds the tongue and nose of the mouse to position the oral cavity, while a second investigator assists by pipetting the bolus into the back of the oral cavity.
6. All instruments and surfaces should be sterilized prior to performing the surgical procedure for targeted intratracheal administration. Preparation of the surgical site should be performed in accordance with your Institutional Animal Care and Use Committee.
7. It is important to minimize animal bleeding during this procedure, so only cut along the midline.
8. The 24-gauge Angiocath is essentially a syringe needle that is inserted into a polymer sheath that is also shaped like a needle with a tapered point. About 1 mm of the needle tip sticks out through the end of the sheath. The Angiocath is packaged with the needle snugly inside the sheath, and both components have independent bases such that the user can easily remove the needle from the sheath (as in **step 9**).
9. This step is time sensitive, as once the Angiocath is inserted into the trachea the delivery of isoflurane will be interrupted. It is therefore important to complete the Angiocath insertion and delivery of *P. aeruginosa* as quickly as possible. To aid in this, it is advised that you preload a pipettor with the 50 μL *P. aeruginosa* bolus prior to the procedure.
10. During the establishment of this technique, it is recommended that a visual marker such as colloidal carbon or India ink be instilled using this method. The marker should consistently and exclusively stain the left lobes of the lung.
11. The conjugate and antigen target of the antibody used for labeling intravascular leukocytes (or a specific subset of leukocytes) can be chosen according to the investigator's specific experimental design and goals. Here, we describe a general labeling of all intravascular leukocytes such that they can be distinguished from leukocytes in the tissue and airspaces.
12. Be sure to include appropriate controls if compensation between detection channels is required for multichannel flow cytometry panels. Also, be sure to remember not to label samples with an antibody conjugated to the same fluorophore used for the intravascular leukocyte label (PE, in this protocol's example).

Acknowledgments

This work was supported by the following awards from the National Institutes of Health: R35GM124911 (C.T.L.) and P20GM121344 Project 2 (B.W.L.). All authors declare no related conflict of interest.

References

1. Lord JM, Midwinter MJ, Chen YF et al (2014) The systemic immune response to trauma: an overview of pathophysiology and treatment. Lancet 384:1455–1465
2. Fagon JY, Chastre J, Vuagnat A et al (1996) Nosocomial pneumonia and mortality among patients in intensive care units. JAMA 275:866–869
3. Jones RN (2010) Microbial etiologies of hospital-acquired bacterial pneumonia and ventilator-associated bacterial pneumonia. Clin Infect Dis 51(Suppl 1):S81–S87
4. Williams BJ, Dehnbostel J, Blackwell TS (2010) Pseudomonas aeruginosa: host defence in lung diseases. Respirology 15:1037–1056
5. Oishi K, Sonoda F, Iwagaki A et al (1993) Therapeutic effects of a human antiflagella monoclonal antibody in a neutropenic murine model of Pseudomonas aeruginosa pneumonia. Antimicrob Agents Chemother 37:164–170
6. Koh AY, Priebe GP, Ray C et al (2009) Inescapable need for neutrophils as mediators of cellular innate immunity to acute Pseudomonas aeruginosa pneumonia. Infect Immun 77:5300–5310
7. Kooguchi K, Hashimoto S, Kobayashi A et al (1998) Role of alveolar macrophages in initiation and regulation of inflammation in Pseudomonas aeruginosa pneumonia. Infect Immun 66:3164–3169
8. Lee K, Cohen JT, Wilson ZS et al (2018) Hemorrhage attenuates neutrophil recruitment in response to secondary respiratory infection by Pseudomonas Aeruginosa. Shock 52(5):506–512
9. Grommes J, Soehnlein O (2011) Contribution of neutrophils to acute lung injury. Mol Med 17:293–307
10. Anderson KG, Mayer-Barber K, Sung H et al (2014) Intravascular staining for discrimination of vascular and tissue leukocytes. Nat Protoc 9:209–222
11. LaBauve AE, Wargo MJ (2012) Growth and laboratory maintenance of Pseudomonas aeruginosa. Curr Protoc Microbiol Chapter 6:Unit 6E.1

Chapter 7

A Mouse Model of Candidiasis

Pilar Fajardo, Ana Cuenda, and Juan José Sanz-Ezquerro

Abstract

The intravenous challenge model of *Candida albicans* infection in mice is a well-established procedure that mirrors disseminated candidiasis in humans. In this model, in which the fungus is delivered into the bloodstream causing a systemic infection, the kidneys are the primary target organs. Mice develop renal failure and septic shock that recapitulates the progressive sepsis seen in humans during severe clinical cases. This model is used to study inflammation and the host immune response against fungal infection. This chapter describes the intravenous candidiasis infection protocol, detailing different steps from the preparation of the inoculum, injection of *Candida*, monitoring of animals, collection of tissue from infected mice, sample preparation and analysis of several parameters related to infection and the inflammatory response.

Key words Candidiasis, Fungal infection, Inflammation, Sepsis, Mouse model, *Candida albicans*

1 Introduction

The fungus *Candida albicans* (*C. albicans*) is a normal component of the human microbiota and is present in mucous surfaces of the gastrointestinal and genital tracts, where it lives as a harmless commensal organism. However, in certain conditions where tissue homeostasis is altered, such as disruption of epithelial barriers or when the host defenses are compromised, *C. albicans* can invade the mucosa and reach the bloodstream. This can cause a systemic infection with potentially lethal consequences, known as candidiasis [1].

In fact, Candida species (and in particular *C. albicans*) are the most frequent fungal agents that cause sepsis due to invasive infections in immunocompromised patients, such as those with AIDS and those undergoing anticancer chemotherapy or organ transplantation. Systemic candidiasis is a real threat that is not fully appreciated and is a serious health problem for patients in critical care units and hospitals. An estimated 400,000 people/year suffer life-threatening infections worldwide and mortality rates associated with candidiasis are high, between 46 and 75% [2, 3]. Moreover,

Wendy E. Walker (ed.), *Sepsis: Methods and Protocols*, Methods in Molecular Biology, vol. 2321,
https://doi.org/10.1007/978-1-0716-1488-4_7,

the appearance of Candida strains resistant to antifungals is on the rise [4]. In order to develop new drugs to efficiently treat this condition, more research is needed to better understand the mechanisms involved in the immune response that normally controls fungal invasion and why they fail in disseminated candidiasis [5].

The use of animal models of disease is required in biomedical research. The complex interactions that take place between the host cells and the pathological agent in different tissues and organs cannot be reproduced faithfully in vitro, in cells in the laboratory. In the case of Candida, the mouse intravenous infection model has been the main protocol used to study systemic candidiasis [6, 7], and this is the procedure we will explain in this chapter [8]. In this model, after direct injection of live *C. albicans* into the bloodstream of mice, a disseminated infection follows and target organs, mainly the kidneys, are affected resembling hyper-inflammatory sepsis [9, 10]. Analyses of mouse survival and organ colonization performed at different time points are used to assess the extent and severity of infection. For instance, histopathological studies can reveal the presence of fungi in the organs (e.g., kidney, spleen, liver) and fungal burden can be quantified in tissue samples by analysis of colony forming units (CFU) in agar-plate culture. Cytometry analyses can be used to reveal which types of cells participate in the immune response. Moreover, genetically modified mouse lines (e.g., knockout for specific genes) can be used to study how particular signaling pathways contribute to disease establishment and progression, by comparing the effects of the candida challenge in wild-type versus mutant animals. Also, the action of different pharmacological inhibitors of signaling cascades can be tested (e.g. kinase inhibitors) [8].

In the following sections we will explain how to prepare the *C. albicans* strain used to infect animals, the protocol for intravenous injection, how to collect samples for biochemical, molecular, cellular, and histological analyses, and the monitoring of mice to assess the outcomes of systemic candidiasis.

2 Materials

2.1 General Equipment and Materials

1. Sterile laminar flow hood (suitable for handling biosafety level 2 materials).
2. Sterile Phosphate buffered saline (PBS): Weigh out 0.256 g of NaH_2PO_4, 1.192 g of $NaHPO_4$, and 8.766 g of NaCl. Transfer to a 1 L graduated cylinder and add water to a volume of 900 mL. Mix and if necessary adjust pH to 6.8 using HCl or NaOH. Make up to 1 L with water. Autoclave the solution.
3. Disposable 1.5–2 mL microcentrifuge tubes.

4. Disposable 15 and 50 mL tubes.
5. 70% ethanol.
6. Disposable barrier pipette tips.
7. Refrigerated benchtop microcentrifuge.
8. Refrigerated centrifuge.
9. Eppendorf shaker.

2.2 Preparation of *C. albicans* Inoculum

1. Commercially available *C. albicans* (strain SC5314) can be maintained on yeast peptone dextrose (YPD) agar plates at 4 °C for up to 1 week. For long-term storage, *C. albicans* are kept in 25% (v/v) glycerol in YPD (1% (w/v) yeast extract, 2% (w/v) peptone, and 2% (w/v) glucose in sterile water) at −80 °C.
2. To prepare YPD agar plates, commercial YPD agar can be used. Alternatively, weigh out 10 g of yeast extract, 20 g of peptone, 20 of glucose, and 15 g of agar, transfer to a 1 L graduated cylinder and add water up to a final volume of 1 L. Heat to boiling while stirring to dissolve the ingredients completely. Autoclave the solution for 15 min at 121 °C. In a sterile laminar flow hood, distribute the solution in plastic Petri dishes (90 mm diameter) until the agar completely covers the plate.
3. Sterilized wire inoculation loop or disposable plastic inoculation loops.
4. Neubauer hemocytometer.

2.3 Infection and Monitoring of Mice

1. Female C57BL/6 mice (available from several supply companies; other mouse strains, such as BALB/c could also be used), 8–12 weeks old (approx. 20–22 g).
2. Weight balance.
3. Heat lamp to warm mouse-tails.
4. Mouse restraint device.
5. Syringe (1 mL) and needles (25 gauge).

2.4 Fungal Load in Kidney

1. Mixer tissue homogenizer.
2. Sterile dissection tools (scissors and forceps).
3. YPD agar plates.

2.5 Flow Cytometry

1. Fluorescently labeled antibodies against immune cell populations of interest (e.g., anti-CD45, -CD3, -CD4, -CD8, -Ly6G, and -F4/80).
2. Roswell Park Memorial Institute medium (RPMI).
3. Solution A (per mouse): 10 mL RPMI, 3% (v/v) fetal bovine serum (FBS), at 37 °C.

4. Solution B (per mouse): 10 mL PBS, 3% (v/v) FBS, 5 mM EDTA at 4 °C.
5. Liberase Solution (per mouse): 2 mL RPMI, 0.2 mg/mL Liberase, 40 mg/mL DNase I.
6. 40 μm Cell Strainer.
7. 3% FBS (v/v) in PBS.
8. Erythrocyte lysis buffer pH 7.4: 4.15 g ammonium chloride (NH_4Cl), 0.5 g sodium bicarbonate ($NaHCO_3$), 20 mg EDTA, 500 mL distilled water, adjust pH to 7.4 using HCl or NaOH.
9. Cell culture plates 96 wells.
10. Dissection tools (scissors and forceps).
11. Neubauer hemocytometer.
12. Trypan blue solution: 0.2 g Trypan blue, 100 mL PBS. Filtrate the solution though 0.22 μm filter.

2.6 Protein Extraction

1. Agate mortar and pestle set.
2. Dry ice and liquid nitrogen.
3. Benchtop vortex mixer (vortex).
4. Lysis buffer: 50 mM Tris–HCl pH 7.5, 1% (v/v) Triton X-100, 1 mM EDTA 1 mM EGTA, 50 mM NaF, 10 mM β-glycerol phosphate, 5 mM sodium pyrophosphate ($Na_4P_2O_7$), 0.27 M sucrose. Add the following reagents just before using the solution: 1 mM sodium orthovanadate (Na_3VO_4), 1 mM benzamidine, 0.1 mM phenylmethyl sulfonylfluoride (PMSF), 0.1% (v/v) 2-mercaptoethanol).

2.7 RNA Extraction

1. Agate mortar and pestle set.
2. Dry ice and liquid nitrogen.
3. Benchtop vortex mixer.
4. RNase-free microcentrifuge tubes 1.5 mL.
5. NZYol reagent for RNA isolation (or similar reagent).
6. Chloroform.
7. 2-Propanol.
8. Ethanol.
9. Microvolume spectrophotometer equipment.

2.8 Histological Analysis of Infected Kidney

1. Formalin solution, neutral buffered 10% for sample fixation.
2. Histology cassettes for embedding and processing samples.

3 Methods

3.1 Preparation of C. albicans Inoculum

1. Using a wire inoculation loop or a disposable plastic inoculation loop, take a small amount of the frozen *C. albicans* glycerol stock and plate (four quadrant streaking) on an YPD agar plate. Incubate at 30 °C for 48 h (*see* **Note 1**).
2. Using an inoculation loop, pick a single *C. albicans* colony and plate it (four quadrant streaking) on an YPD agar plate. Incubate at 30 °C for 48 h.
3. Using an inoculation loop, pick a single *C. albicans* colony and resuspend it in 1 mL of sterile PBS in a 1.5 mL microcentrifuge tube. From now on this suspension will be called the "initial solution."
4. Determine the number of cells in the initial solution using a Neubauer hemocytometer. Dilute the initial solution 1:100 in sterile PBS, count the colony forming units (CFU) and then calculate the cell concentration (CFU per mL) present in the initial solution (*see* **Note 2**).
5. Calculate the number of CFUs needed for the experiment and take the volume of initial solution necessary to prepare the total inoculum (diluted in PBS). In our experience, the number of CFUs needed to induce candidiasis is 1×10^5 CFUs per mouse and the maximum injection volume is 200 μL. For example, if the number of mice used in the experiment is 20, then the total CFUs needed are 1×10^5 (CFU/mouse) $\times$ 20 (mice) $= 2 \times 10^6$ CFUs and the total volume is 20 (mice) $\times$ 0.2 (mL/mouse) = 4 mL (*see* **Note 3**).

3.2 Infection of Mice

1. The day before the infection, arrange mice into control and experimental groups (*see* **Note 4**). The control group will be inoculated with PBS and the experimental group with *C. albicans* (*see* **Note 5**).
2. Mix the *C. albicans* inoculum, by inverting the tube, and load in a 1 mL syringe.
3. Restrain the mouse in the restraint device, hold the tail and warm it with the lamp. Locate the lateral vein by slightly rotating the tail. Disinfect the injection site with 70% ethanol and insert the needle into the vein at a slight angle (*see* **Note 6**).
4. Inject 200 μL of PBS (control mice) or *C. albicans* inoculum (experimental group) into the lateral tail vein.
5. Make sure that the injection is correct (you should not encounter resistance when inserting the needle or when pressing the plunger). The lumen of the vein becomes clear when the inoculum is injected and then filled again with blood. After inoculation, remove the needle and press the injection site with a tissue.

3.3 Mouse Monitoring

1. Monitor the mice at least once a day for any signs of discomfort or disease symptoms. Examine each mouse's mobility, food and water intake, temperature, breathing, posture/position, and hair condition (*see* **Note 7**). Also, weigh the mice daily as a readout of their health state. If a mouse has lost 25% (or more) of the initial weight, or if it shows signs of severe illness, it should be removed from the cage and humanely sacrificed, according to ethical regulations in animal experimentation (e.g., by cervical dislocation or CO_2 asphyxiation). More than 25% weight loss is an indicator of impending death.

3.4 Fungal Load

All the procedures from now on will be carried out in a laminar flow hood in order to reduce the risk of contamination.

1. Immediately after death, or after euthanasia at experimentally determined time points, spray the mouse with 70% ethanol, open it with dissection tools and remove the kidneys aseptically (*see* **Note 8**). Extract the kidney as cleanly as possible, eliminating all adipose tissue (fat).
2. Place the kidney in a 1.5 mL microcentrifuge tube and weight it on a precision scale, since the total amount of CFU per kidney is normalized to the weight of the organ. Rinse the organ with 1 mL of sterile PBS.
3. Disaggregate the organs in 1 mL sterile PBS using a mixer tissue homogenizer. Clean the homogenizer with 70% ethanol and with PBS after the homogenization of each kidney.
4. Make serial dilutions (e.g., 1:10; 1:100) of the stock homogenates in sterile PBS.
5. Plate 10 μL of the stock homogenates and 10 μL of each dilution on YPD agar plates that have been previously marked with three divisions (Fig. 1), one per each dilution, and extend using an inoculation loop. Plate always from more diluted to less diluted solutions. Incubate at 30 °C for 24 h to allow the *C. albicans* to grow. After the indicated time, count the colonies and calculate the CFU per mL of homogenate. Fungal load is expressed as CFU per gram of kidney taking into consideration the dilutions and the weight of the organ.

3.5 Flow Cytometry Analysis of Immune Cell Infiltration

1. Extract the kidney as indicated (*see* Subheading 3.4), and cut it into small pieces with a scalpel on a petri dish with RPMI. Collect all the pieces into a 15 mL tube and digest the kidney pieces with 2 mL of Liberase solution for 20 min at 37 °C on a shaking platform.
2. Stop the reaction with 10 mL of solution A, filter the suspension through 40 μm cell strainers (one per mouse) and add 1 mL of solution A to wash the filter.

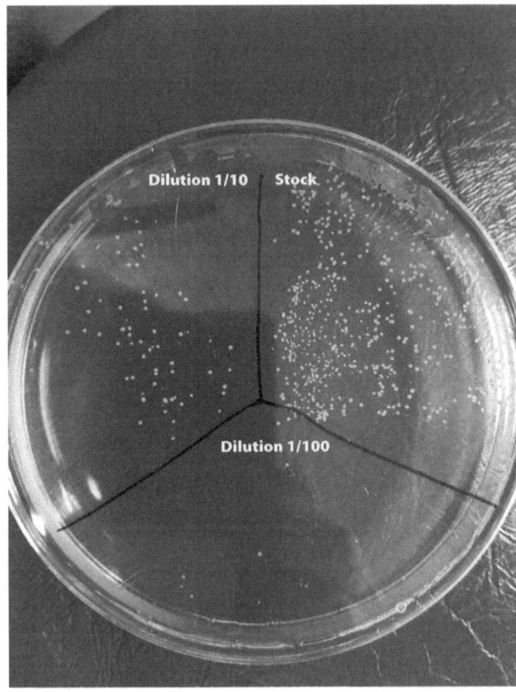

Fig. 1 *C. albicans* colonies on YPD agar plates. Serial dilutions are indicated

3. Centrifuge at 538 rcf for 5 min at 4 °C. Aspirate supernatants and wash two times with 10 mL solution B (centrifuging at 538 rcf for 5 min at 4 °C).
4. Remove erythrocytes, by adding 1.5 mL of Erythrocyte lysis buffer for 2 min at room temperature, and stop the reaction with 5 mL of PBS 3% FBS. Centrifuge at 538 rcf for 5 min at 4 °C.
5. Aspirate the supernatants and resuspend by shaking the pellet in 1 mL of PBS 3% FBS.
6. Count the cells with a Neubauer hemocytometer making a 1:2 dilution with "Trypan Blue Solution" in order to count only the living cells (*see* **Note 2**).
7. In each flow cytometry sample, use approximately 1×10^6 cells. Also use 1×10^6 cells to set up the cytometer conditions for each antibody, using the appropriate positive and negative controls (e.g., as a positive control, use (for each antibody) single stained cells that express the specific marker, and as negative control use unstained cells). Calculate the volume (μL) of cell suspension needed to have 1×10^6 cells and plate it into a 96-well round-bottom plate.
8. Centrifuge the plate at 290 rcf for 5 min at 4 °C.
9. Stain the cells with the desired combination of fluorescently labeled antibodies against cell surface markers for different immune cell populations (e.g., anti-CD45, -CD4, -CD8,

-Ly6G, and -F4/80). Prepare the antibody mixes needed and add the mix to the corresponding wells. Resuspend the pellet by pipetting and incubate for 20 min at 4 °C.

10. Analyze the labeled cells in a flow cytometer following standard protocols for fluorescence activated cell sorting (*see* **Note 9**).

3.6 Protein Extraction

1. Extract the kidney as indicated (*see* Subheading 3.4), and snap-freeze it in liquid nitrogen. The organ or pieces of tissue can be stored for months at −80 °C, until ready to use.
2. The frozen tissues are subsequently pulverized in liquid nitrogen using a mortar, until a fine powder is obtained. Place the powder in a microcentrifuge tube (*see* **Note 10**).
3. For protein extraction, add threefold volume excess of ice-cold Lysis buffer to the powdered tissue and homogenize on ice using a vortex. Centrifuge the homogenates at 20,780 rcf for 15 min at 4 °C to remove insoluble material. Collect the supernatant and measure protein concentration using a standard colorimetric assay (e.g., Bradford). Snap-freeze the lysates in aliquots in liquid nitrogen and store at −80 °C (*see* **Note 11**).

3.7 RNA Extraction

1. Extract the kidney as indicated above (*see* Subheading 3.4) and process tissue as described for protein (*see* Subheading 3.6, **steps 1** and **2**). Put the obtained powder in an RNase-free microcentrifuge tube (*see* **Note 12**).
2. Add 1 mL of NZYol reagent per sample and homogenize with a vortex.
3. Add 200 μL of chloroform per sample, cover the sample tightly and vortex until homogenized. Allow the sample to stand for 15 min at room temperature.
4. Centrifuge at 20,780 rcf for 15 min at 4 °C. Transfer the aqueous phase to a new RNase-free tube, add 500 μL of 2-propanol per sample and mix by gently inverting the tube ten times. Leave at −20 °C at least 2 h or overnight.
5. Centrifuge at 20,780 rcf for 15 min at 4 °C. The RNA precipitate will form a pellet on the bottom of the tube.
6. Remove the supernatant and wash the RNA pellet by adding 1 mL of cold 75% ethanol. Centrifuge at 20,780 rcf for 15 min at 4 °C. Repeat the process twice.
7. Leave the pellet to dry and add 20–50 μL of RNase-free water. Incubate the samples at 60 °C for 10 min at 300 rpm on a shaker.
8. Measure the RNA concentration with a microvolume spectrophotometer.

3.8 Histological Analysis

1. Extract the kidney as indicated (*see* Subheading 3.4), place it into a histology cassette and fix the samples in formalin for 24 h at room temperature. After 24 h, change the cassettes from formalin to PBS.
2. Embed the tissue in paraffin. Cut the paraffin tissue blocks in 5 μm sections using a microtome (follow standard procedures for histological preparations).
3. Stain sections as desired to visualize changes in the histological structure of the organ, presence of fungus or infiltrated immune cells. You can use standard histological staining (e.g., periodic acid Schiff (PAS) and hematoxylin-eosin) to visualize the presence of Candida hyphae in the kidney, as an indicator of infection (Fig. 2) (*see* **Note 13**).

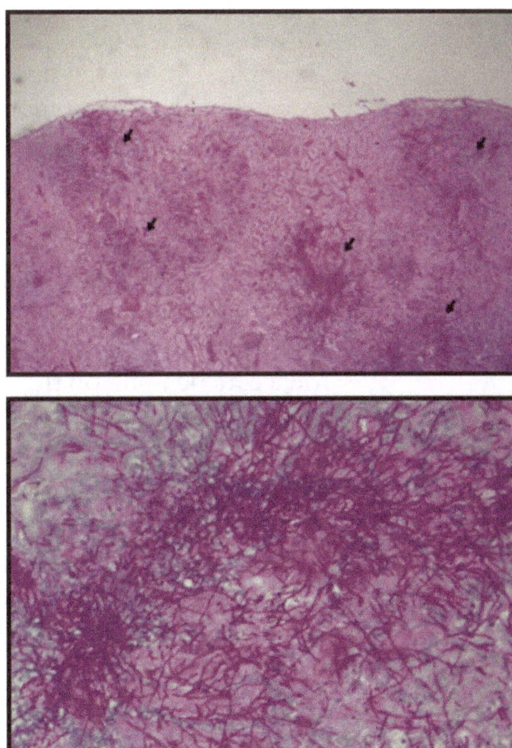

Fig. 2 Representative PAS-hematoxylin–eosin staining of kidney sections from mice at day 3 postinfection. Black arrows in the upper panel indicate *C. albicans* growth areas (×5 magnification). Bottom panel is a high-magnification of one *C. albicans* growth area (×20 magnification)

4 Notes

1. Carry out all the procedures at room temperature in a laminar flow hood. Working with *C. albicans* requires the use of a biosafety level 2 laboratory.

2. To count the number of Candida CFU using a Neubauer chamber, add 10 μL of 1:100 dilution of the initial solution into the chamber. Visualize through the microscope the counting grid. The grid has nine square subdivisions; count the cells present in the bigger squares (the ones at the corners). Start counting the cells in the first square and continue with the remaining ones. At the end, the average of the four squares will be the total cell number. When you calculate the CFUs in the initial solution, you have to take into account the dilution (1:100) and the Neubauer factor ($\times 10^4$).

3. Take into account that you need to prepare *C. albicans* solution in excess, since there is always a significant volume loss. For example if your experiment has 20 mice and you need 4 mL, you will make 7 mL (enough to inoculate 35 mice).

4. Infection of mice with *C. albicans* is a procedure classified as severe in animal experimentation; ensure that you comply with the ethical and welfare guidelines in place at your institution/country and that you have all the necessary permissions and certificates required to work with experimental animals. Also, make sure you use a method to calculate the adequate sample size for your experiments, such as power analysis. In our experience, depending on the particular experiment, between 6 and 8 animals are needed in each group [11].

5. The transportation of the *C. albicans* inoculum to the animal facility should be done in the adequate packaging for biosafety level 2 organisms.

6. Intraperitoneal (IP) injection is another route for the induction of systemic infection. However, in this model, higher fungal inoculum is required [8, 12].

7. In survival experiments, mice should be monitored till day 30 after *C. albicans* inoculation.

8. In *C. albicans* systemic infection the kidney is the main target organ, but other organs such as spleen, liver and brain can be also analyzed for fungal load. Also, the collected organs can be analyzed for other parameters: histological changes, immune cell population analysis, cytokine and chemokine profile, signaling protein activation and gene expression.

9. Many examples of this type of analysis can be found in the literature, e.g [8, 9, 13–15].

10. Both protein and RNA can be extracted from one kidney, using half of the powder to obtain protein and the other half to extract RNA, allowing different analyses to be performed on the same organ.
11. The obtained protein extracts can be used to analyse the status of different signaling pathways by Western blot (e.g., activation by phosphorylation), study the differential expression of proteins between samples (e.g. by antibody arrays) or determine the profile of cytokines or chemokines in the tissue.
12. RNA is extracted from the samples to study gene expression changes associated with infection, for example using qPCR or RNAseq.
13. You can use standard histological staining (e.g., periodic acid Schiff (PAS) and hematoxylin–eosin) to visualize the presence of Candida in the kidney, as a qualitative readout of infection.

Acknowledgments

Research in the laboratory of JJS-E and AC is supported by a grant from MINECO (SAF2016-79792-R (AEI/FEDER, UE)). PF holds a MINECO FPI fellowship. We thank Dr. D. Alsina-Beauchamp for help with the kidney images.

References

1. Wisplinghoff H, Seifert H, Wenzel RP et al (2006) Inflammatory response and clinical course of adult patients with nosocomial bloodstream infections caused by Candida spp. Clin Microbiol Infect 12:170–177
2. Brown GD, Denning DW, Gow NA et al (2012) Hidden killers: human fungal infections. Sci Transl Med 4:165rv13
3. Kullberg BJ, Arendrup MC (2015) Invasive Candidiasis. N Engl J Med 373:1445–1456
4. Kim JY (2016) Human fungal pathogens: why should we learn? J Microbiol 54:145–148
5. Netea MG, Joosten LA, van der Meer JW et al (2015) Immune defence against Candida fungal infections. Nat Rev Immunol 15:630–642
6. Szabo EK, MacCallum DM (2011) The contribution of mouse models to our understanding of systemic candidiasis. FEMS Microbiol Lett 320:1–8
7. Luttich A, Brunke S, Hube B et al (2013) Serial passaging of Candida albicans in systemic murine infection suggests that the wild type strain SC5314 is well adapted to the murine kidney. PLoS One 8:e64482
8. Alsina-Beauchamp D, Escos A, Fajardo P et al (2018) Myeloid cell deficiency of p38gamma/p38delta protects against candidiasis and regulates antifungal immunity. EMBO Mol Med 10: e8485
9. Lionakis MS, Lim JK, Lee CC et al (2011) Organ-specific innate immune responses in a mouse model of invasive candidiasis. J Innate Immun 3:180–199
10. Lionakis MS, Swamydas M, Fischer BG et al (2013) CX3CR1-dependent renal macrophage survival promotes Candida control and host survival. J Clin Invest 123:5035–5051
11. Charan J, Kantharia ND (2013) How to calculate sample size in animal studies? J Pharmacol Pharmacother 4:303–306
12. Cheng S, Clancy CJ, Xu W et al (2013) Profiling of Candida albicans gene expression during intra-abdominal candidiasis identifies biologic processes involved in pathogenesis. J Infect Dis 208:1529–1537
13. Drummond RA, Wallace C, Reid DM et al (2014) Cutting edge: failure of antigen-specific CD4+ T cell recruitment to the kidney during

systemic candidiasis. J Immunol 193:5381–5385

14. Ngo LY, Kasahara S, Kumasaka DK et al (2014) Inflammatory monocytes mediate early and organ-specific innate defense during systemic candidiasis. J Infect Dis 209:109–119
15. Yu YR, O'Koren EG, Hotten DF et al (2016) A protocol for the comprehensive flow cytometric analysis of immune cells in normal and inflamed murine non-lymphoid tissues. PLoS One 11:e0150606

Chapter 8

Francisella tularensis Infection of Mice as a Model of Sepsis

Charles T. Spencer, Mireya G. Ramos Muniz, Nicole R. Setzu, and Michelle A. Sanchez Guillen

Abstract

Sepsis results from the dysregulated immune response to infection. While the stimulator and progression of the septic response is poorly understood, the systemic production of a storm of cytokines is common in all etiologies of sepsis. While the complexity of this uncontrolled cascade is difficult to replicate using single molecule agonist, for example, lipopolysaccharide (LPS), several whole organism models can stimulate this cytokine storm. Herein, we detail protocols developed to trigger and analyze the systemic septic response in mouse models using the bacterium *Francisella tularensis*.

Key words Francisella tularensis, Sepsis, Mouse model of sepsis, Bacterial model of sepsis

1 Introduction

In response to infection, a localized immune response is initiated by cells of the innate immune system to eliminate, or at least control, the invader. These cells possess a number of pathogen recognition receptors (PRRs) that recognize a variety of pathogen-associated molecular patterns (PAMPs) contained on infectious microbes and the damage-associated molecular patterns (DAMPs) released from host cells due to microbial infection. PRR engagement by the PAMP or DAMP initiates an intracellular signaling cascade, the exact details of which vary based upon the PRR(s) activated, and result in the production of IL-1β and subsequent initiation of the inflammatory response. The autocrine and paracrine signaling response to IL-1β triggers the production of a flood of cytokine signals which are sent back and forth between local innate immune cells. This results in the activation of these local effector cells, recruitment of additional cells to the site of infection, and migration of dendritic cells to draining lymph nodes to activate adaptive immune cells. Once the microbial invader is eliminated and the PRR is no longer engaged, regulatory cells counter the inflammatory signals shutting down the inflammatory response.

Wendy E. Walker (ed.), *Sepsis: Methods and Protocols*, Methods in Molecular Biology, vol. 2321,
https://doi.org/10.1007/978-1-0716-1488-4_8,

For the majority of microbes, this immune reaction remains localized to the site of infection and is sufficient to maintain the health of the host. For reasons currently unknown, this inflammatory response can become dysregulated leading to an overwhelming systemic response to infection, termed sepsis. The clinical definition of sepsis has evolved over the decades and now can be "simply" defined as a dysregulated immune response subsequent to infection leading to multiorgan system failure. This ends up classifying sepsis as a syndrome with no singular cause or progression to disease. Regardless, sepsis can be fatal in up to 20% of cases and has the highest incidence of disease worldwide [1, 2].

Despite the prevalence of sepsis, relatively little is known about the pathophysiology of the disease. Currently, sepsis is diagnosed using the Sequential (or Sepsis-related) Organ Failure Assessment (SOFA) score, which evaluates the functioning of six organ systems: respiration, coagulation, liver, cardiovascular, central nervous system, and renal. However, dysfunction in only two of these systems is sufficient for a diagnosis of sepsis resulting in 30 different progressions to sepsis from a single trigger [3]. In addition, it is unclear why some microbes have a higher likelihood of triggering the septic response or under what genetic or environmental conditions the septic response becomes more likely. Indeed, dozens of different genetic polymorphisms have been associated with progression to sepsis [4–7].

The dysregulation of the immune response to infection called sepsis results in a cascade of inflammatory and anti-inflammatory cytokines and chemokines produced by immune cells. While some key players are known, IL-1β, IL-6, TNF-α, IFN-γ, and so on [8–11], there is no defined pattern of cytokine production that can be specifically associated with sepsis. This means that the "cytokine storm" induced by different microbes leading to sepsis could involve different players. Indeed, attempts to directly reduce the levels of these cytokines as treatment options have failed to prevent sepsis [12]. Since sepsis results from this dysregulated immune response to an infection, it is dependent upon the presence of infectious microbes. Likewise, there have been no properties identified that specifically indicate whether or not a particular microbe may trigger a septic response.

A diverse collection of infectious organisms are known to cause disease by triggering a dysregulation of inflammation culminating in the septic cytokine storm [13–23]. To simplify the study of the septic response, intravenous administration of lipopolysaccharide (LPS) has been used to activate the PRRs and trigger the systemic septic response. While successful, this reductionist approach triggers only a single PRR and its signaling pathway. Intact organisms contain numerous PAMPs capable of activating multiple PRRs and multiple signaling pathways which, when integrated, result in a cytokine storm that is vastly different from LPS administration alone. Therefore, recent experiments into the cause, progression, and treatment of sepsis have utilized specific whole organisms to provide this multifactorial response.

Infection with *Francisella tularensis,* which causes the human disease tularemia, is accompanied by a large increase in various cytokine levels in the blood/serum. It is thought that this large increase in cytokines is the direct cause of tularemia pathology [13, 14]. Indeed, mice that succumb to *F. tularensis* infection display high levels of serum cytokines compared with those that survive infection, despite similar levels of bacterial burden [24]. In other words, infected patients die from an overactive immune response against the *F. tularensis* bacteria rather than from products of the bacteria itself. During the progression of tularemia, multiple organ systems become dysregulated leading to increased SOFA scores, diagnostic for sepsis. This makes *F. tularensis* infection a reliable and biologically relevant stimulator of the septic response.

The species *F. tularensis* comprises four subspecies: *F. tularensis* subsp. *tularensis*, *F. tularensis* subsp. *holarctica*, *F. tularensis* subsp. *novicida*, and *F. tularensis* subsp. *mediasiatica* [25]. These subspecies vary in their global distribution and importantly in their ability to infect and cause disease in humans and mice. *F. tularensis* subsp. *tularensis* (type A) is the most virulent of the four subspecies and is predominately found in North America. It is classified as a risk group 3 select agent requiring high containment and security protocols for its use. *F. tularensis* subsp. *holarctica* (type B) is less virulent, still causing human infections though rarely death, and is primarily found in Europe and Asia. Because of its reduced ability to cause disease, it is classified as a risk group 2 agent and can be used under normal laboratory conditions. *F. tularensis* subsp. *novicida* appears to be avirulent in humans, even putting its ability to infect humans in question, and, as such, is considered a risk group 1 organism. Little is known about *F. tularensis* subsp. *mediasiatica* except that it is predominantly present in central Asia.

In addition to human infection, the highly virulent *F. tularensis* subsp. *tularensis* recapitulates the severity of human disease in mouse models of infection [25, 26]. This strain is desirable for mouse studies because of its high relevance to human disease; however, due to the need for BSL3 level containment and special permission from the CDC, much of our understanding of *F. tularensis*-induced immunity comes from studies of the highly attenuated *F. tularensis* subsp. *holarctica*-derived live vaccine strain (LVS) or the mouse pathogen *F. tularensis* subsp. *novicida* [25, 27]. Although *F. tularensis* subsp. *holarctica* and *F. tularensis* subsp. *novicida* cause little or no apparent disease in humans, both can nonetheless cause lethal tularemia-like disease in mice [25, 26]. For all of these bacterial subspecies, infection of mice produces a cytokine storm in a dose-dependent manner, allowing for studies aimed at investigating initiation of the septic response, as well as progression of the cytokine cascade and involvement of multiorgan system failure utilizing conspecific controls. However, the virulence, as measured by the lethal dose at which half the mice develop severe disease (LD_{50}), varies not only between subspecies but also by route of administration [28].

Detailed herein are protocols developed by the authors for the experimental use of *F. tularensis* to trigger the septic response and the methods by which that response is analyzed. These protocols are written for (A)BSL1 or (A)BSL2 level laboratories, although modifications to these procedures would allow for similar use in BSL3 level laboratories.

2 Materials

2.1 Bacteria

1. *Francisella tularensis* subsp. *novicida* Utah 112 (BEI Resources, catalog #NR-13)—BSL1.
2. *Francisella tularensis* subsp. *holarctica* CDC LVS (BEI Resources, catalog #NR-646)—BSL2.
3. *Francisella tularensis* subsp. *tularensis* SCHU S4 (BEI Resources, catalog #NR-643)—BSL3.

2.2 Mice

1. C57BL/6 J mice (*see* **Note 1**).

2.3 Equipment

1. 1 L plastic beaker.
2. Digital scale.
3. Digital infrared thermometer.
4. Petri dishes.
5. Cell scrappers.
6. 48 well plate.
7. Micropipettes and micropipette tips.
8. Heparinized capillary tubes.
9. Microcentrifuge tubes size 1.5 mL.
10. Microcentrifuge tubes size 600 μL.
11. Needles and syringes.
12. Anesthesia vaporizer similar to Veterinary Anesthesia Systems cat# V3000PS (recommended for intramuscular and intradermal administration routes, optional for the subcutaneous, intraperitoneal, intravenous, and intranasal administration routes).
13. Mouse restraint device (optional for the intravenous administration route), similar to ProLab Supply cat# MH-100.

14. Emersion grinder, similar to Tissue Tearor (BioSpec Products, cat# 985370).
15. Capillary tubes.
16. Multiplex ELISA kit selected according to the needs of your study and budget. We recommend MilliPlex Mouse Cytokine/Chemokine Magnetic Bead Panel—Premixed 32 Plex (Millipore Sigma, cat# MCYTMAG-70K-PX32) for ease of use and broad coverage.
17. 96-well plates.
18. Adhesive plate sealers.
19. Titer plate shaker.
20. Handheld magnetic separator block (Millipore Sigma, cat# 40-285).
21. MAGPIX plate reader with xPONENT software (Luminex corporation), or similar Luminex plate reader.

2.4 Reagents

1. Chocolate Agar II plates.
2. 5% Sheep's Blood Agar (SBA) plates.
3. Muller Hinton (MH) Broth Powder.
4. Sodium chloride (NaCl).
5. Proteose Peptone Tryptone.
6. Agar.
7. Defibrinated sheep's blood.
8. IsoVitaleX Enrichment Media (Becton Dickinson, cat# 211875).
9. Glycerol.
10. PBS.
11. Milliplex Drive Fluid.
12. Isoflurane, or other approved anesthesia.
13. Cavicide, Virkon or other approved disinfectant.
14. 70% Ethanol.
15. RNA stabilization reagent, such as RNAlater.

3 Methods

3.1 Preparation of 5% Sheep's Blood Agar (SBA) Plates

1. Add the intended volume of distilled water to a media flask or bottle.
2. Weight out the appropriate amounts of MH Broth, NaCl, tryptone, and agar powders (Table 1) and add to the water. *Do not add the IsoVitaleX or sheep's blood yet.*

Table 1
Recipe to Prepare 5% Sheep's Blood Agar (SBA) Plates

Total volume	MHBroth	NaCl	Tryptone	Agar	Isovitalex	Sheep's blood
1000 mL	21 g	5 g	10 g	16 g	10 mL	50 mL
500 mL	10.5 g	2.5 g	5 g	8 g	5 mL	25 mL
250 mL	5.25 g	1.25 g	2.5 g	4 g	2.5 mL	12.5 mL
100 mL	2.63 g	0.63 g	1.25 g	2 g	1.25 mL	6.25 mL

3. Mix the ingredients on a stir plate and autoclave at 121 °C for 15–20 min.
4. Once complete, transfer to a 56 °C water bath to cool (*see* **Note 2**).
5. Add reconstituted IsoVitaleX to the mixture (*see* **Note 3** and Table 1).
6. Add defibrinated sheep's blood to yield a final concentration of 5% blood.
7. Dispense ~17–25 mL of broth per petri dish (*see* **Note 4** and Table 1).
8. Allow the plates to solidify and fully cool before covering and inverting.
9. Store the plates inverted in a sealed container at 4 °C for up to 2–3 months, plates should be discarded when dehydration and cracking are evident.

3.2 Preparation of Cryopreservation Buffer

1. Prepare cryopreservation buffer for freezing bacteria as a 15% glycerol in 1× PBS solution.
2. Sterilize the solution by autoclaving or filtering (*see* **Note 5**).

3.3 Sub-Culturing Francisella Tularensis

Subspecies of *F. tularensis* vary in their growth properties and nutritional requirements. In general, they are considered fastidious organisms, though *Francisella tularensis* subsp. *novicida* can be grown on standard LB agar plates. The growth of *Francisella tularensis* subsp. *holarctica* or *Francisella tularensis* subsp. *tularensis* requires the addition of nutritional supplements. This is generally provided by the addition of the BBL IsoVitaleX supplement, which contains a number of additional vitamins, amino acids, coenzymes, dextrose, ferric ions, and NAD. However, supplementation of LB with L-Cysteine can facilitate the growth of *Francisella tularensis* subsp. *holarctica* though its virulence can be attenuated during prolonged culture; therefore, we do not recommend its sole use. In our experience, *Francisella tularensis* subsp. *holarctica* and *Francisella tularensis* subsp. *tularensis* require heme/iron in order to maintain their virulence when grown in culture.

Likewise, the growth rate of the strains varies with *Francisella tularensis* subsp. *novicida* doubling every ~1.5 h while *Francisella tularensis* subsp. *holarctica* and *Francisella tularensis* subsp. *tularensis* double every 4–5 h [29]. It is therefore critical that sufficient time is given for colonies to grow. Depending upon environmental conditions and bacterial health this can be anywhere from 24–48 h. We recommend checking growth plates at 24 h; but if small colonies are present, leave the plates for another day for anything other than a simple colony count.

1. Use proper aseptic technique at all times. To increase sterility the following procedures can be done in the biosafety cabinet.
2. Streak or spread an aliquot from the stock bacteria culture onto a sheep blood agar (SBA) or chocolate agar plate (*see* **Note 6**).
3. Place the SBA plates in the 37 °C incubator for 24–48 h until large bacterial colonies have formed. Do not harvest while the colonies are still small.
4. Add 2 mL of cryopreservation buffer to the plate.
5. With a plate scraper gently scrape all the bacteria from the plate to the edge of the petri dish.
6. Collect all the cryopreservation buffer containing the bacteria into a 15 mL sterile tube, using a 1 mL pipette. Pipette repeatedly to disrupt cell clumps.
7. Repeat **steps 4–6** with each plate. To increase the final concentration, the cryopreservation buffer containing the bacteria can be transferred to subsequent growth plates for scraping.
8. Once all of the bacteria have been harvested, vortex the final tube to evenly distribute the bacteria.
9. Aliquot 120 μL of the bacteria into microcentrifuge tubes.
10. Every 3–4 microcentrifuge aliquots, vortex the mixture for ~7 s to redistribute the bacteria (*see* **Note 7**).
11. Freeze the aliquoted bacteria at −80 °C overnight (*see* **Note 8**).
12. Randomly select 3–4 aliquots from the frozen stock to determine the CFU for the lot using the titration method (*see* **Note 9** and Subheading 3.4).
13. Apply this CFU count to the entire lot. This should be redetermined at least every 3 months for a given lot and every time a new lot is made.

3.4 Determining the Francisella Tularensis Titer

1. Using a 48 well flat-bottomed plate dispense 900 μl of 1× sterile PBS into 8 wells (*see* **Note 10**).
2. Add 100 μL of the bacterial stock to the first well and mix well by repeated pipetting, to create a 1:10 dilution in well #1.

3. Create a tenfold serial dilution series by transferring 100 μL from the first well into the 2nd well and mix thoroughly by pipetting. Continue through all wells; the last well does not need to have 100 μL removed from it.
4. Plate 100 μL from the last four wells (wells 5, 6, 7, 8) onto separate SBA plates and spread evenly with a sterile bacterial spreader (*see* **Note 10**).
5. Place all plates into the 37 °C incubator for 48 h (*see* **Note 11**).
6. After bacterial colonies have grown, count the number of colonies on the plates and calculate the starting concentration using the following formula:

$$\text{Colony count} \times \text{Dilution factor}^{\text{Well number}}/\text{Volume plated (in mL)}$$

For example, following the procedure above (tenfold serial dilution) and obtaining a count of 50 bacterial colonies on the plate from the 6th dilution well would result in the following calculation: $50 \times 10^6/0.1\ \text{mL} = 5 \times 10^8$ bacteria/mL.

3.5 Infection Procedures

F. tularensis is naturally acquired through multiple modes of transmission, including: aerosol, direct contact with infected animals, eating or drinking contaminated food or water, and transmission by arthropod vectors (e.g., ticks, biting flies, fleas, and mosquitoes) [25]. While the other transmission routes are globally ubiquitous, there appears to be a geographic dependence for arthropod vector transmission. Mosquito-based transmission is currently limited to northern Europe while tick-based transmission is dominant in North America with biting flies transmitting in limited regions of the far mid-west USA.

The route of transmission is critical to consider as the signs, symptoms, and progression of the disease vary based on how *F. tularensis* enters the body [26]. There are six main forms of tularemia defined in humans: ulceroglandular, glandular, oculoglandular, oropharyngeal, typhoidal, and pneumonic. Ulceroglandular and glandular tularemia are the most common forms of tularemia and are usually caused by the bite of a tick or deer fly. Oculoglandular and oropharyngeal tularemia are infection of the eye or mouth/throat, respectively, and occur through exposure to contaminated food or water or through contact with infected animals. Typhoidal tularemia incorporates a combination of general symptoms without specific localized symptoms indicative of the other manifestations. Pneumonic tularemia is the most serious and often lethal form of tularemia and results from breathing aerosol droplets containing the bacteria; it can also develop from spread of bacteria from the bloodstream to the lungs, if other manifestations are left untreated. Despite these numerous forms, all can result in the development of a septic response, yet the exact mechanism leading to sepsis currently remains unclear.

Table 2
Injection route volumes and doses for *Francisella tularensis* subspecies

Route		Subcutaneous (s.c.)	Intramuscular (i.m.)	Intradermal (i.d.)	Intraperitoneal (i.p.)	Intravenous (i.v.)	Intranasal (i.n.)
Injection volume (μL)		100–1000	10–50	10–50	100–1000	100–500	35–50
Spp. dose (cfu)	*tularensis*	10	10	10	10	10	10
	LVS	10^6	10^3	10^6	10	10	10^3
	novicida	10^5	10^2	10^5	10	10	10

The selection of a route for laboratory-based administration of the bacteria should, therefore, be relevant to the model being tested. Intradermal and subcutaneous administration of the bacteria can mimic an arthropod-based transmission route. Intraperitoneal and intravenous administration can mimic the acquisition of typhoidal tularemia. Oral gavage replicates consumption of contaminated food or water and development of oropharyngeal tularemia. Intranasal instillation mimics aerosol delivery of contaminated droplets, though is not as penetrant as aerosol exposure, which to date has proven challenging.

Each route of administration has its own challenges, proponents, and dosing requirements [28]. Due to its aggressiveness, the dosage of *Francisella tularensis* subsp. *tularensis* used in mouse studies is ~10 cfu, which results in a 50% lethal dose (LD_{50}), independent of route of administration (Table 2). *Francisella tularensis* subsp. *holarctica* LVS is more attenuated, and therefore, provides a larger range of doses for different routes of administration. While systemic administration by intravenous or intraperitoneal administration is still highly aggressive (LD_{50} ~ 10 cfu), intranasal instillation requires ~10^3 cfu to reach a LD_{50} while intradermal and subcutaneous administration requires ~10^6 cfu. This larger range makes it easier to manage variation between injections and allows for the analysis of a dose-dependent response. *Francisella tularensis* subsp. *novicida* is more adapted to infecting mice and, therefore, its LD_{50} dose is generally a log lower. Oral gavage, though technically possible, is rarely used as it has a high degree of variability, even amongst mice within a single experiment. This is thought to be due to variations in the survival of bacteria in the stomach. Due to this variability, oral gavage will not be described herein.

3.6 Subcutaneous (s.c.) Administration (Figs. 1 and 2)

Subcutaneous administration can be performed without anesthesia in a conscious animal, using a single-handed restraint. The rate of absorption is lower than that from intraperitoneal and intramuscular injections. Subcutaneous administrations are made into the loose skin over the interscapular or inguinal area. Injection volumes by this route are highly flexible as you are injecting into loose connective tissue and range from 100–1000 μL for adult mice.

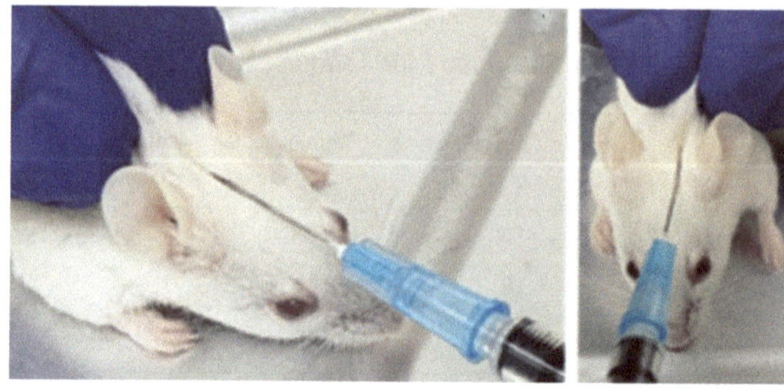

Fig. 1 Subcutaneous administrations over the interscapular area. With a restrained and/or anesthetized mouse, tent the skin over the scapula with your thumb and forefinger and gently penetrate the skin with the needle

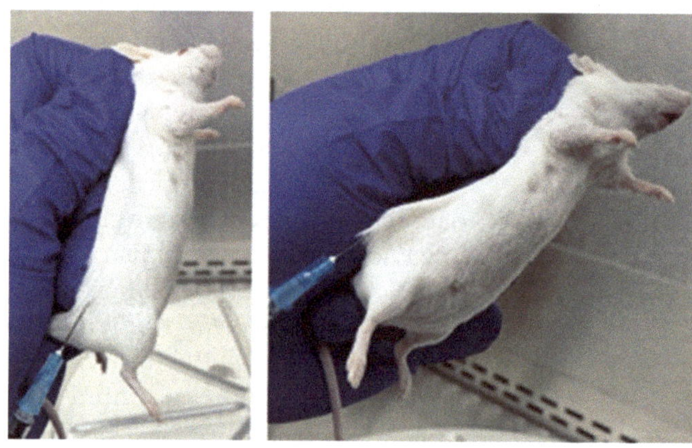

Fig. 2 Subcutaneous administrations over the inguinal area. With a restrained and/or anesthetized mouse, scruff the mouse and gently penetrate the skin over the flank with the needle

1. Load the bacteria into a syringe attached to a 25–30G needle.
2. For subcutaneous administrations over the interscapular area, manually restrain the mouse and place on a clean, solid surface (*see* **Note 12** and Fig. 1). Using the thumb and forefinger, tent the skin by gently pinching the skin above each scapula. Gently penetrate the skin of the interscapular area with the needle and inject.
3. For subcutaneous administrations over the inguinal area, manually restrain the mouse by single hand scruffing restraint (*see* **Note 12** and Fig. 2). Insert the needle just under the skin overlying the mouse's thigh and parallel to the body line. Advance the needle a few millimeters through the subcutaneous tissue and inject.
4. Return the mouse to its cage.

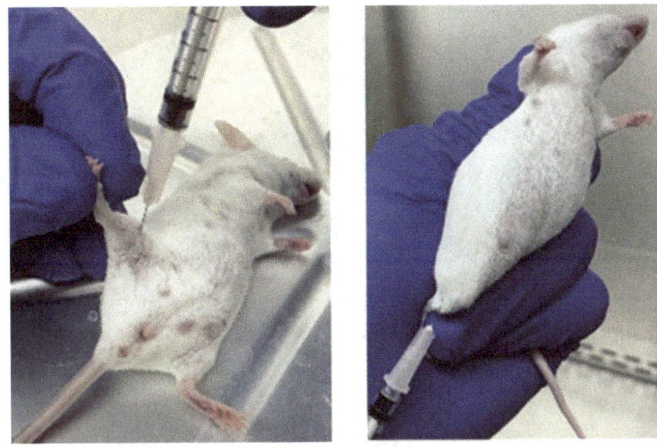

Fig. 3 Intramuscular injection. With an anesthetized mouse, extend the hind leg and penetrate the muscle with the needle parallel to the femur. Alternatively, scruff the mouse and trap the hind leg and penetrate the muscle with the needle parallel to the femur

3.7 Intramuscular (i.m.) Injection (Fig. 3)

Intramuscular administration can be done without analgesia in a conscious mouse using a single-handed restraint. However, the leg manipulations require a dexterous hand; it is much simpler to anesthetize the mouse. As you are injecting directly into the thigh muscle, the injection volume is limited to 10–50 μL for adult mice.

1. Load the bacteria into a syringe attached to a 27–30G × ½ in. needle.
2. Anesthetize the mouse and lay it on its side (*see* **Note 13**).
3. Stretch the hind leg to allow access to the muscle of the thigh.
4. Insert the needle into the thigh muscle and carefully inject.
5. Return the mouse to its cage and monitor until it regains consciousness.

3.8 Intradermal (i.d.) Administration (Fig. 4)

Intradermal administration is a delicate procedure as the needle must penetrate the epidermis but not the subcutaneous layer of the skin. The thin skin of the mouse requires a fine needle to be used, 25–30G. As you are distending the skin with this method, the maximum injection volume by this route is 50 μL for adult mice. Because of the delicate maneuvers of this route, it is strongly recommended that the mouse be anesthetized to prevent it moving during the injection.

1. Load the bacteria into a syringe attached to a 25–30G × ½ in. needle.
2. Anesthetize the mouse.
3. Clip or otherwise remove a patch of fur from the area on the flank above the hind quarters (*see* **Note 14**).

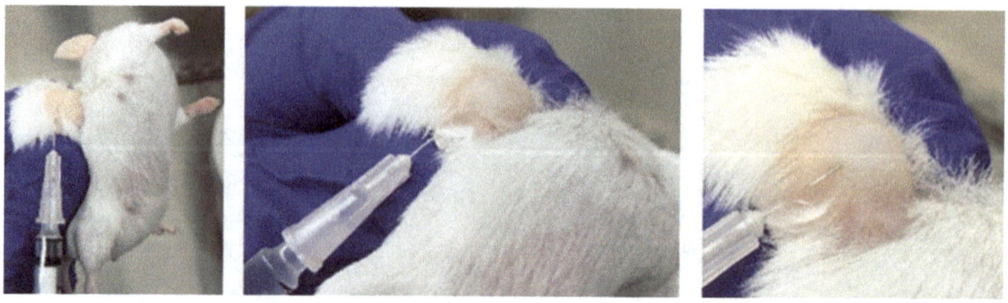

Fig. 4 Intradermal administration. With an anesthetized mouse, drape the skin over your finger and gently penetrate the epidermis with the needle

4. Wipe the exposed area with 70% ethanol on a gauze sponge or swab.
5. Drape the skin over your outstretched index finger and hold firmly with your thumb and middle finger so that the exposed skin is positioned at the highest point and is taught.
6. Lay the point of the needle bevel side up and perpendicular to your finger. It should be angled so as to pass over top of your finger should you puncture through the skin.
7. Slowly advance the needle just under the superficial layer of epidermis about 2 mm into the skin.
8. Once the needle has penetrated the skin, relax the tension on the skin and inject.
9. A bleb or welt should appear in the tented skin at the site of injection. If this bump does not form, the injection has penetrated the dermis into the subcutaneous tissue.
10. Return the mouse to its cage and monitor until it regains consciousness.

3.9 Intraperitoneal (i.p.) Administration (Fig. 5)

Intraperitoneal administration allows for a long period of absorption: fluid is absorbed from the peritoneal cavity between one-half and one-fourth as rapidly as from the intravenous route. A larger volume of isotonic solution can be administered by this route. This route delivers the most flexibility for the injection volume, which can range between 50 μL–1 mL. If approved by the local IACUC, up to 3 mL is tolerable by the mouse, although this large volume is not recommended.

1. Load the bacteria into a syringe attached to a 25–30 G needle.
2. Manually restrain the mouse by single hand scruffing restraint in a supine position with its posterior end slightly elevated (*see* **Note 12**). With the mouse in a head-down position, the internal organs will slide slightly toward the animal's head, helping to prevent accidental penetration by the needle.

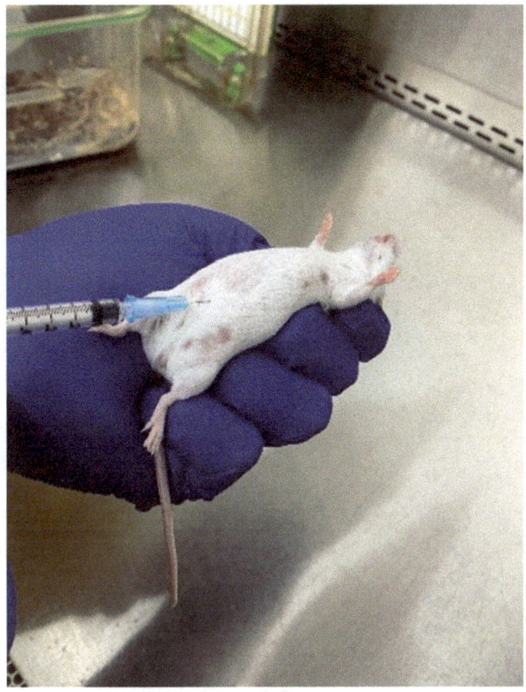

Fig. 5 Intraperitoneal administration. With a restrained and/or anesthetized mouse, scruff the mouse and gently penetrate into the abdominal cavity

3. The needle and syringe should be kept nearly parallel to the animal's spine in order to avoid accidental penetration of the viscera.
4. Insert the needle through the abdominal wall with a ~15° angle between the needle and abdominal surface.
5. Make sure that the needle is far enough into the peritoneal cavity to prevent subcutaneous delivery, but not so far as to puncture the viscera. Inject the contents of the syringe.
6. Return the mouse to its cage.

3.10 Intravenous (i.v.) Injection (Fig. 6)

Intravenous administration is the most rapid method for systemic delivery of a substance into the mouse; however, it is also the most challenging. While this procedure can be done on a conscious animal, anesthetizing or restraining the mouse by a manner other than scruffing is highly recommended. This allows for the use of both hands during the injection; one to provide tension and control of the tail and the other to guide the needle. Mastering this procedure requires considerable practice. It is recommended to warm the whole body or tail to dilate the blood vessel and provide an easier target. The goal with this administration route is to thread the needle into the lateral tail vein, which is not much larger than the needle itself.

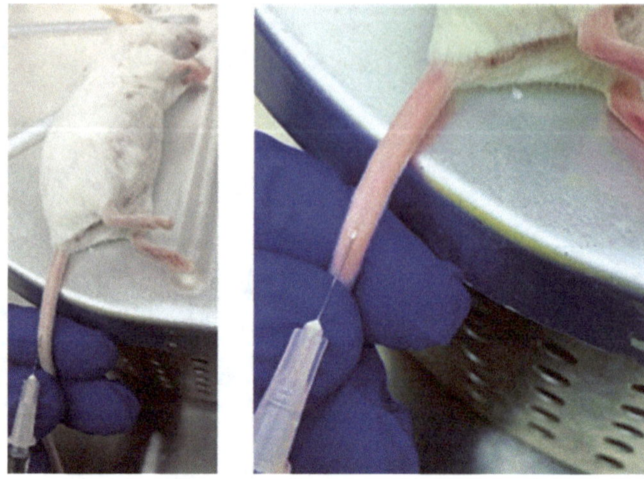

Fig. 6 Intravenous injection. With a restrained and/or anesthetized mouse, drape the tail over the edge of a box and slide the needle into the lateral tail vein

1. Load the bacteria into a syringe with a 30 G × ½ in. needle.
2. Anesthetize the mouse and lay it in a supine position on an elevated platform with the tail loosely draped off the platform or table edge.
3. Alternatively, manually restrain the mouse by scruffing and anchoring the tail and hind leg between the small finger and palm.
4. Lay the point of the needle bevel side up so that it is parallel with the lateral tail vein.
5. Carefully advance the needle into the tail and thread it into the tail vein.
6. Gently push on the plunger. If no resistance is met, proceed to inject. Resistance means that you are in the tissue and not in the vein. Remove the needle and try again higher up the tail, more proximal to the body.
7. Return the mouse to its cage and, if anesthesia was used, monitor until it regains consciousness.

3.11 Intranasal (i.n.) Instillation (Fig. 7)

Intranasal instillation provides an easy method for delivery of substances into the lungs. Due to the droplet size, this method primarily delivers substances into the upper airways of the lungs and will not completely fill the lungs as would result from an aerosol exposure. This route requires that the mouse inhale a droplet while breathing; to reduce the chances of aspiration pneumonia, the liquid should be applied via small sequential droplets until the entire volume has been applied. The volume of administration by this route is 35–50 μL.

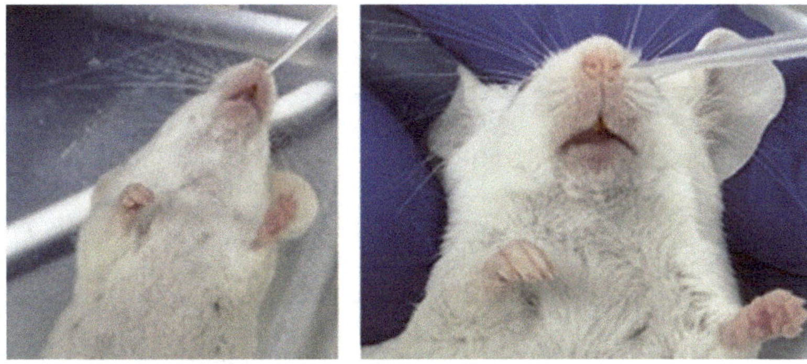

Fig. 7 Intranasal instillation. With a restrained and/or anesthetized mouse, use a pipette tip to place a drop on the nostril

1. Anesthetize the mouse and lay it in a supine position with its nose positioned adjacent to a nose cone, this will maintain anesthesia while leaving the nose accessible for droplet administration.
2. Alternatively, manually restrain the mouse by scruffing and anchoring the tail between the small finger and palm. The mouse should be held in a supine position with its head elevated.
3. Place the end of a micropipette at the external nares and slowly dispense the solution. The recommended volume is 10 μL at a time per nostril (*see* **Note 15**).
4. Return the mouse to its cage and, if anesthesia has been used, monitor until it regains consciousness.

3.12 Assessment Monitoring of Health

Mice challenged with *F. tularensis* should be observed visually frequently in their cage as their posture and behavior provides important information as to their disease response. In the first 48–72 h, the mice will be motile, have normal posture and will interact with others, if cohabitating. When opening the cage under the hood, one will notice that healthy mice are alert and active. Also, a healthy mouse's fur and eyes will be clean, a sign of normal grooming. When restraining a mouse, its posture will be normal and the animal will be active. When mice are placed in a novel environment, a healthy mouse will defecate or urinate; this shows that the mouse is healthy with a normal appetite.

Following inoculation, infected mice should be closely monitored throughout the course of infection for signs of wasting and fur loss. Humane endpoints for mice infected with *F. tularensis* generally include a weight-loss maximum, inactivity/lethargy, respiratory distress, conditions affecting eating and drinking ability, temperature change, and severe clinical signs of infection. Weight-loss maxima generally range from 15–20% of initial body weight. Our observation has been that a 15% weight loss is too low to distinguish between mice that will recover and those that are unable to do so. Contrary to sick humans, mice tend to become hypothermic following infection which can be monitored internally or externally [30]. Experimentally infected mice reaching the humane endpoint should be euthanized immediately by lab personnel per procedures established in an approved IACUC protocol. With clear justification, IACUC may allow death as an endpoint.

Typically, at 72 h postinfection, mice will begin to look lethargic, have reduced motility and a hunched posture. Also, sick mice will be isolated from the others in a shared cage. The mouse's fur will be ruffled and will have a greasy appearance either because the mouse has stopped grooming entirely, or is not doing this often enough. The eyes will be semi-closed and will have a white crust around them; these mice should be monitored closely as death is near. Corneal damage, seen as a white layer of material at the center of the eyes reminiscent of cataracts, are also not an uncommon occurrence, especially when blood is being drawn every other day. When restrained, the sick animals will have a hunched posture and will not resist handling. When mice are placed in a novel environment, the sick mouse will not defecate or urinate, a sign of poor or lost appetite. When restraining sick mice their body will be softer if they are hydrated, in contrast they will be slightly dry with opaque fur if they are not hydrated.

Mice should be monitored, at least daily, for clinical signs (CS) of illness/disease and scored according to the severity of their symptoms. We have found that during days 3–5 postinoculation, the condition of the mice can change very rapidly and monitoring should be increased to 3–4 times per day. We have generated a scale for the CS where: 0 = healthy, 1 = slight ruffled fur, 2 = increase ruffled fur or hunched posture, 3 = both severely ruffled fur and hunched posture, and 4 = lethargic/immobile [24]. We have noticed that mice in the CS = 3–4 category will sometimes develop a white mucoid discharge from the eye. Mice in this state should be monitored closely as they are, generally, within hours of death.

3.13 Monitoring Weight loss (Fig. 8, See Note 16)

1. Disinfect and sanitize the biosafety cabinet surface (*see* **Note 17**).
2. Inside the biosafety cabinet, place the scale and a 1 L plastic beaker to contain mice while weighing. Sanitize these items with 70% ethanol.
3. Transfer a cage of mice to the biosafety cabinet.
4. Remove the cage lid and observe the mice's behavior for clinical signs of disease.
5. Place the beaker on top of the scale and zero the scale.

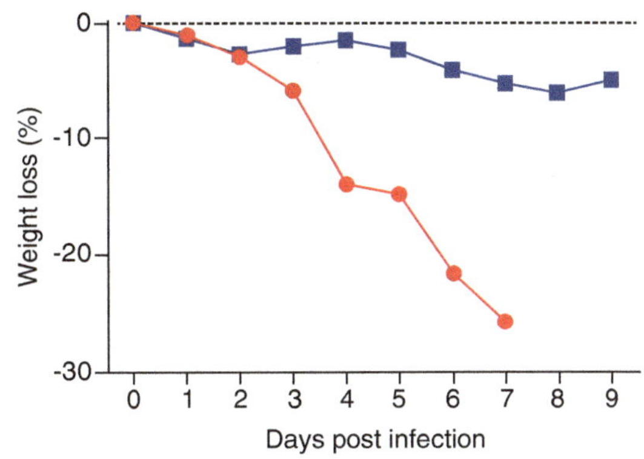

Fig. 8 Weight loss following *F. tularensis* infection. Representative graph showing weight change as the percent change from initial body weight of a group of animals following infection

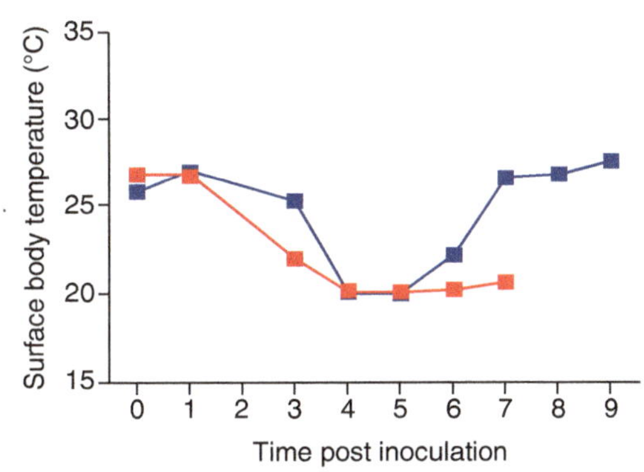

Fig. 9 Surface temperature change following *F. tularensis* infection. Representative graph showing surface body temperature as measured with infrared thermometer of a group of animals following infection

6. Lift a mouse by the tip of the tail and place it in the beaker.
7. Allow the scale to stabilize as much as possible and record the weight. The scale will fluctuate as the mouse moves.
8. Gently return the mouse to the cage and the cage to the rack.

3.14 Monitoring Body Temperature (Fig. 9)

1. Disinfect and sanitize the biosafety cabinet surface (*see* **Note 17**).
2. Inside the biosafety cabinet, place the infrared thermometer. Sanitize the thermometer with 70% ethanol.
3. Transfer a cage of mice to the biosafety cabinet.
4. Remove the cage lid and observe the mice's behavior for clinical signs of disease.

5. While the mouse remains in the cage, measure the surface body temperature using the noncontact infrared thermometer and record for each mouse.
6. Return the cage to the rack.

3.15 Sample Collection

As a model of sepsis, it is critical to monitor the systemic immune response to *F. tularensis* infection and not just the local tissue response. Our protocol monitors the systemic response through measurements of the cytokine storm in blood as well as the local immune response within organs [24]. However, since the organ must be removed for analysis, this only provides a single time point for those numbers. Through careful procedures, blood samples can be obtained every other day to monitor the kinetics of the cytokine storm developing in response to *F. tularensis* infection. Special care must be taken when obtaining frequent blood samples that the total volume does not drop too low such that the mouse goes into hypovolemic shock. To prevent this, blood collection is limited to 1.5% of the mouse's body weight over the 14-day time course of *F. tularensis* infection. To control the amount of blood taken every other day, we recommend drawing from the retro-orbital sinus with a capillary tube. In our opinion, this allows for more precise control of the amount of blood withdrawn at any one time as blood flow stops almost immediate when the mouse is released. Light pressure can be applied to the closed eye to ensure rapid bleeding cessation.

3.16 Retro-Orbital Blood Collection (Fig. 10, See Note 18)

1. Anesthetize the mouse and grasp the skin behind the head, slightly stretching the facial skin.
2. Position a heparinized capillary tube behind the eyeball at the anterior of the eye socket.
3. Apply very light pressure and roll the tube back and forth to rupture the sinus. Do NOT advance the capillary. Keep it gently pressed against the back of the eye socket to allow the blood to collect. If too much pressure is applied, the capillary can pierce the brain.
4. Collect ~40 μL of blood from infected mice into the heparinized capillary tube and transfer blood to a 1.5 mL microcentrifuge tube by gravity drainage or a gentle puff of air.
5. Centrifuge the blood samples at 1000 × *g* for 10 min at 4 °C.
6. Transfer the plasma to a new microcentrifuge tube.
7. Store plasma samples at −80 °C for future analysis of the systemic cytokine storm (*see* Subheadings 3.19–3.21).

3.17 Euthanasia (See Note 19)

There are several methods of euthanasia approved by the American Veterinary Medical Association (AVMA), which most IACUC offices follow. Exceptions to these can be considered by most IACUC boards with reasonable scientific justification. The most

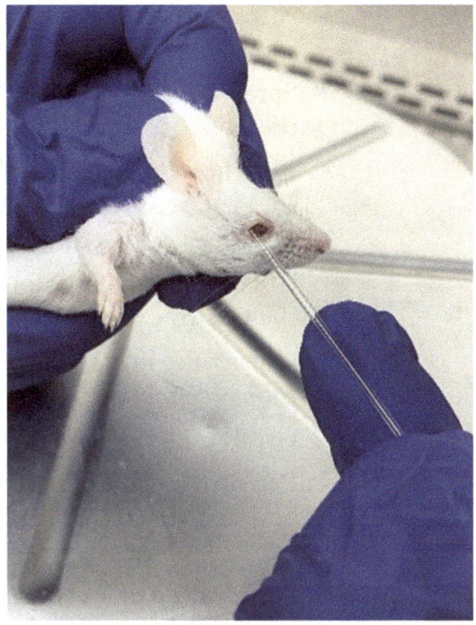

Fig. 10 Retro-orbital blood collection. With an anesthetized mouse, use a capillary to gently rupture the retroorbital sinus

common methods of euthanasia include carbon dioxide overdose and anesthesia (e.g., isoflurane) overdose. A secondary method of confirming death must also be performed; these include, cervical dislocation, bilateral thoracotomy, decapitation, or removal of vital organs (detailed herein).

1. Transfer the mouse to be euthanized into a fresh cage or induction chamber (*see* **Note 20**).
2. Transfer the cage of mice to be euthanized to the procedure room (*see* **Note 21**).
3. Place the euthanasia cage in the carbon dioxide (CO_2) chamber and turn on the gas following AVMA guidelines for flow rate.
4. Observe the mice for loss of consciousness and cessation of respiration.
5. Maintain CO_2 flow rate for at least 1 min after the mice stop breathing.
6. Turn off the CO_2 flow and remove the cage from the chamber.
7. Perform an approved secondary method of euthanasia as approved by IACUC (such as vital organ removal, *see* Subheading 3.18).
8. Dispose of the carcasses per institutional standard operating procedures.

3.18 Tissue Collection

1. After euthanasia (*see* Subheading 3.17), place the mouse in a supine position.
2. Spray the mouse with 70% ethanol to disinfect the surface of the skin.
3. Lift the skin over the abdomen and make a shallow cut with surgical scissors so as not to cut the organs, especially the intestines as this can result in contamination.
4. Create parallel cuts along both side of the chest cavity through the ribs to expose the thorax.
5. Collect the liver, lung and spleen from infected mice into 5 mL tubes containing 1 mL PBS.
6. Grind each tissue with a mechanized immersion grinder.
7. Perform serial dilutions and plating to quantify bacterial load, as described in Subheading 3.4.
8. Centrifuge the remaining tissue homogenate at $500 \times g$ for 10 min to pellet cellular material.
9. Remove the supernatant and store at −80 °C for future analysis of the local cytokine storm, as described in Subheadings 3.19–3.21.
10. Resuspend the tissue homogenate in an RNA stabilization reagent and store at −80 °C for future analysis. RNA harvested from this homogenate can be used for real-time RT-PCR analysis of gene expression.

3.19 Analysis of the Cytokine Storm

As discussed, the septic response is an unregulated, systemic inflammatory response with no single etiology or manifestation. Therefore, it is important to consider the totality of the inflammatory response, as much as possible. Due to the limited plasma blood volume obtained from repeated collection (*see* Subheading 3.16), two possibilities exist for analysis, pooling the plasma samples or multiplexing the analytes. While pooling plasma samples provides ease of analysis, it reduces the number of samples within each group. Multiplexing, though more complicated, increases the power of analysis between each group [24]. Recent advances in multiplex ELISA analyzed by flow cytometry (e.g., Luminex) allows for the interrogation of 1–100 analytes, most from a very small sample, e.g., the 25 μL plasma resulting from the procedure in Subheading 3.16. These multiplexed ELISA systems generally use beads, each specific for a particular analyte, which can be combined into a single analysis stream. Described herein is the procedure used by our laboratory for this complex analysis (*see* **Note 22**). In general, multiplex ELISA kits will be supplied with materials sufficient to quantify 96 samples (1 plate). For the kits routinely used by our laboratory (*see* **Note 22**), we have found that there is sufficient material to quantify 120 samples per kit with no reduction in sensitivity. As this is a fluorescence-based assay, standards do not need to be repeated on the second plate since slight fluctuations in kinetics has little effect on fluorescence. Additional plates for these extra samples can often be purchased directly, or prior plates can be cleaned and reused.

3.20 Preparation of Multiplex ELISA Buffers

1. Dilute the wash buffer from the kit as instructed. Let the buffer warm up to room temperature. Mix to bring all the salts into solution.
2. In a sterile flask, dilute the 10× wash buffer to 1× wash buffer, by diluting 60 mL of the 10× wash buffer into 540 mL of deionized water.
3. Quality controls 1 and 2 come in powder. Reconstitute both of the quality controls by adding 250 μL of deionized water. Once the water is added, close the vial and vortex for 30 s, make sure that all the powder has gone into solution. Put the vials on ice for 10 min in order to fully hydrated.
4. Standards come in powder, reconstitute them by adding 250 μL of deionized water. Once water is added close the vial and vortex for 30 s, make sure that all the powder has gone into solution. Put the vials on ice for 10 min in order to fully resuspend. This is "standard 6."
5. Add 1 mL of deionized water to the lyophilized serum matrix (*see* **Note 23**). Resuspend well and allow the matrix to reconstitute for 10 min sitting on ice.

3.21 Luminex Assay

1. Prepare the beads. For premixed bead sets: sonicate the bottle for 30 s then vortex for 1 min before use. For individual vials of beads: sonicate each vial for 30 s then vortex for 1 min. Add 60 μL of each antibody bead into a mixing bottle or tube. Bring the final volume of the mixing bottle to 3 mL with assay buffer. This volume is sufficient for 120 samples (*see* **Note 24**). Vortex the mixed beads once again for at least 1 min.
2. Allow the plasma samples to thaw to room temperature (20–25 °C).
3. Using a multichannel pipette, add 200 μL of wash buffer into each well of the plate.
4. Cover the plate using an adhesive plate sealer.
5. Shake the plate for 10 min at room temperature on the microtiter plate shaker.
6. Once the incubation on the shaker is done, decant the wash buffer into the sink, invert the plate and gently tap onto paper towels to remove the residual buffer.
7. Add 25 μL of each standard in the designated wells (*see* **Note 25**).

8. Add 25 μL of quality controls in the designated wells (*see* **Note 25**).
9. Add 25 μL of assay buffer to the background well.
10. Add 25 μL serum matrix to background, standards and control wells only (*see* **Note 23**).
11. Using a multichannel pipette, add 25 μL assay buffer into all the experimental sample wells.
12. Add 25 μL of plasma sample into the appropriate experimental sample wells.
13. Vortex the mixed bead vial and then add 20 μL to each well. Place the adhesive seal back on and wrap the plate with foil (*see* **Note 26**).
14. Shake the plate on the microtiter plate shaker overnight at 4 °C protected from light.
15. The next day, stop the shaker and bring the plate to room temperature.
16. Warm the included detection antibody bottle to room temperature.
17. For magnetic beads, place the plate on a hand-held magnet (*see* **Note 27**).
18. Leave the plate on the magnet for 60 sec. This step draws the beads to settle at the bottom of the plate. *Do NOT shorten this step or you may lose the beads in the subsequent step.*
19. Leaving the plate in the magnet, gently decant contents into the sink and remove most of the residual liquid by inverting the plate onto paper towels and lightly dabbing the plate. It is NOT desirable to remove all of the liquid as for a traditional ELISA.
20. Remove the plate from the magnet.
21. Using a multichannel pipette, add 200 μL wash buffer into each well of the plate.
22. Put the adhesive seal back on, wrap with foil and shake the plate for 30 sec.
23. Repeat washing **steps 17–20** two additional times.
24. Using a multichannel pipette, add 25 μL of the detection antibodies into each well of the plate.
25. Seal the plate with the adhesive seal, wrap in foil and shake the plate for 1 h at room temperature.
26. Remove plate from shaker. DO NOT DECANT.
27. Using a multichannel pipette, add 25 μL streptavidin-phycoerythrin to each well already containing the detection antibodies.

28. Seal the plate with the adhesive seal, wrap in foil and shake the plate for 30 min at room temperature.
29. Repeat washing **steps 17–22** three times.
30. Using a multichannel pipette, add 150 μL of Drive fluid to all wells.
31. Seal the plate with the adhesive seal, wrap in foil and shake the plate for 5 min at room temperature.
32. Remove the plate and run the plate on MAGPIX with xPONENT software for analysis (*see* **Note 28**).

4 Notes

1. All mouse procedures should be approved by your Institutional Animal Care and Use Committee (IACUC) prior to experimentation involving mice. It is important that you select a strain that is appropriate for your experimental system. The most common strains used in sepsis research include BALB/c and C57BL/6. It should be noted that these strains have different responses to infection with C57BL/6 being more inflammatory than BALB/c. Thus, the authors recommend the use of the C57BL/6 strain.
2. It is critical to cool the temperature of the agar below 65 °C prior to addition of the IsoVitaleX, glucose, and sheep's blood.
3. Cysteine can be supplemented for IsoVitaleX for the LVS strain, at 1 g/L.
4. This volume can be adjusted as desired; however, do not make the agar too thin or it will tear during use.
5. Glycerol and PBS can also be sterilized prior to mixing and combined under sterile conditions.
6. The agar plates should be at room temperature. You can use multiple plates depending on the concentration desired.
7. The bacteria tends to settle in the tube so if this step is skipped the concentrations will not be similar in all aliquots.
8. The bacteria should be left in the freezer for at least 24 h before determining the titer.
9. Following this procedure, the variance between aliquots is <5% in our hands.
10. This may need to be adjusted depending on the concentration of your bacteria preparation.
11. It is best to check the CFU on the plate at the 24-h time point to observe the progress and growth of the bacterial colonies.
12. This procedure is easily done on a conscious mouse, however, the mouse can also be anesthetized if desired.

13. Alternatively, manually restrain the mouse by scruffing and anchor the tail between the small finger and palm and the hind leg between the fourth finger and small finger.
14. Clipping or removing the fur is not required for this procedure but increases visibility of the injection area and is highly recommended while training. Once proficient, the procedure can be done by wetting the area with 70% ethanol and brushing the fur aside to reveal the skin.
15. Excess volume or rapid injection may provoke suffocation and death.
16. For accurate monitoring, it is recommended to weigh the animals at the same time every day in order to avoid artifacts from the natural daily fluctuations in body weight.
17. This process varies by institution but generally involves the use of one or more disinfectants (e.g., 10% bleach, Cavicide®, Virkon®, and/or 70% ethanol).
18. For the last time point before euthanasia, a cardiac puncture can be performed to obtain the blood sample.
19. Methods of euthanasia vary and should be approved by individual IACUC review. Always follow the procedures set in place by the IACUC.
20. It is recommended that a specific box be dedicated for euthanasia to prevent the smell of death from affecting the remaining animals. Remaining animals should also NOT witness the euthanasia.
21. It is recommended that animals NOT be euthanized in holding rooms as this can increase the anxiety of all the other animals.
22. In our experience, Millipore MilliPlex® assays provide consistent and reliable data. This procedure generally follows the manufacturer's recommendations.
23. Serum matrix is only used when testing serum or plasma samples.
24. Calculation example: if using 20 bead vials, 60 μL will be withdrawn from each vial, resulting in a total bead volume of 1.2 mL. 1.8 mL of assay buffer must therefore be added to bring the final volume up to 3 mL.
25. Before adding standards and controls to the wells, vortex each vial.
26. This step is critical. The beads are light sensitive, therefore you must protect the bead vial and the plate (once beads are added) from light. Also, vortex the beads bottle at least every three wells to prevent the beads from settling to the bottom of the bottle.

27. Make sure the safety pins on both sides of the magnet are released, so that the plate can sit completely on the magnet. Once the plate is in the right position, push the pins back on to secure the plate.
28. Plates can be stored at 4 °C but should be analyzed within 24 h. Shake the plate for 5 min at room temperature prior to analysis.

Acknowledgments

Research for which these protocols were developed was supported by *the National Institute of Allergy and Infectious Diseases* of the National Institutes of Health under award number SC1 AI148753 *to CTS.*

References

1. Fleischmann C, Scherag A, Adhikari NKJ et al (2016) Assessment of global incidence and mortality of hospital-treated sepsis. Current estimates and limitations. Am J Respir Crit Care Med 193:259–272
2. Adhikari NKJ, Fowler RA, Bhagwanjee S et al (2010) Critical care and the global burden of critical illness in adults. Lancet 376:1339–1346
3. Chousterman BG, Swirski FK, Weber GF (2017) Cytokine storm and sepsis disease pathogenesis. Semin Immunopathol 39:517–528
4. Sutherland AM, Walley KR (2009) Bench-to-bedside review: association of genetic variation with sepsis. Crit Care 13:210
5. Arcaroli J, Fessler MB, Abraham E (2005) Genetic polymorphisms and sepsis. Shock 24:300–312
6. Texereau J, Pene F, Chiche J-D et al (2004) Importance of hemostatic gene polymorphisms for susceptibility to and outcome of severe sepsis. Crit Care Med 32:S313–S319
7. Chung LP, Waterer GW (2011) Genetic predisposition to respiratory infection and sepsis. Crit Rev Clin Lab Sci 48:250–268
8. Mera S, Tatulescu D, Cismaru C et al (2011) Multiplex cytokine profiling in patients with sepsis. APMIS 119:155–163
9. Gouel-Chéron A, Allaouchiche B, Guignant C et al (2012) Early interleukin-6 and slope of monocyte human leukocyte antigen-DR: a powerful association to predict the development of sepsis after major trauma. PLoS One 7:e33095
10. Wu H-P, Chen C-K, Chung K et al (2009) Serial cytokine levels in patients with severe sepsis. Inflamm Res 58:385–393
11. Kellum JA, Kong L, Fink MP et al (2007) Understanding the inflammatory cytokine response in pneumonia and sepsis: results of the genetic and inflammatory markers of sepsis (GenIMS) study. JAMA Intern Med 167:1655–1663
12. Opal SM, Dellinger RP, Vincent J-L et al (2014) The next generation of sepsis clinical trial designs: what is next after the demise of recombinant human activated protein C?*. Crit Care Med 42:1714–1721
13. Mares CA, Ojeda SS, Morris EG et al (2008) Initial delay in the immune response to Francisella tularensis is followed by hypercytokinemia characteristic of severe sepsis and correlating with upregulation and release of damage-associated molecular patterns. Infect Immun 76:3001–3010
14. Sharma J, Mares CA, Li Q et al (2011) Features of sepsis caused by pulmonary infection with Francisella tularensis type a strain. Microb Pathog 51:39–47
15. Wauquier N, Becquart P, Padilla C et al (2010) Human fatal Zaire Ebola virus infection is associated with an aberrant innate immunity and with massive lymphocyte apoptosis. PLoS Negl Trop Dis 4:e837
16. Villinger F, Rollin PE, Brar SS et al (1999) Markedly elevated levels of interferon (IFN)-γ, IFN-α, interleukin (IL)-2, IL-10, and tumor necrosis factor-α associated with fatal Ebola virus infection. J Infect Dis 179:S188–S191

17. Huang R, Zhang L, Gu Q et al (2014) Profiles of acute cytokine and antibody responses in patients infected with avian influenza a H7N9. PLoS One 9:e101788
18. Bondu V, Schrader R, Gawinowicz M et al (2015) Elevated cytokines, thrombin and PAI-1 in severe HCPS patients due to sin Nombre virus. Viruses 7:559
19. Figueiredo LT, Souza WM, Ferres M et al (2014) Hantaviruses and cardiopulmonary syndrome in South America. Virus Res 187:43–54
20. Khaiboullina SF, Rizvanov AA, Lombardi VC et al (2013) Andes-virus-induced cytokine storm is partially suppressed by ribavirin. Antivir Ther 18:575–584
21. Gralinski LE, Baric RS (2015) Molecular pathology of emerging coronavirus infections. J Pathol 235:185–195
22. Huang K-J, Su I-J, Theron M et al (2005) An interferon-γ-related cytokine storm in SARS patients. J Med Virol 75:185–194
23. Zhou J, Chu H, Li C et al (2014) Active replication of middle East respiratory syndrome coronavirus and aberrant induction of inflammatory cytokines and chemokines in human macrophages: implications for pathogenesis. J Infect Dis 209:1331–1342
24. Mireya G, Ramos-Muniz MP, Sanchez MA, Setzu N, Portillo PS, Garza KM, Gosselink KL, Spencer CT (2018) Obesity exacerbates the cytokine storm elicited by Francisella tularensis infection of females and is associated with increased mortality. BioMed Res Int 2018:3412732
25. Sjostedt A (2007) Tularemia: history, epidemiology, pathogen physiology, and clinical manifestations. Ann N Y Acad Sci 1105:1–29
26. Penn RL (2000) Francisella tularensis (Tularemia). In: Mandell GL, Bennett JE, Dolin R (eds) Principles and practice of infectious diseases. Churchill Livingstone, Philadelphia, PA, pp 2674–2685
27. Tarnvik A, Eriksson M, Sandstrom G et al (1992) Francisella tularensis—a model for studies of the immune response to intracellular bacteria in man. Immunology 76:349–354
28. Cowley S, Elkins K (2011) Immunity to Francisella. Front Microbiol 2:26
29. Morris BJ, Buse HY, Adcock NJ et al (2017) A novel broth medium for enhanced growth of Francisella tularensis. Lett Appl Microbiol 64:394–400
30. Mei J, Riedel N, Grittner U et al (2018) Body temperature measurement in mice during acute illness: implantable temperature transponder versus surface infrared thermometry. Sci Rep 8:3526

Chapter 9

A Mouse Model of Necrotizing Enterocolitis

Belgacem Mihi, Wyatt E. Lanik, Qingqing Gong, and Misty Good

Abstract

Necrotizing enterocolitis (NEC) is an acute inflammatory disease that unforeseeably develops in very low birth weight premature infants. NEC is characterized by impairment of the intestinal barrier resulting in intestinal necrosis and multisystem organ failure. Animal models of NEC have contributed significantly to a better understanding of the underlying molecular mechanisms of the disease and facilitated the exploration of potential new therapeutic strategies. Here, we provide a detailed protocol that recapitulates some of the main histological and transcriptional features of human NEC in newborn mice.

Key words Necrotizing enterocolitis, Neonatal mice, Mucosal barrier, Intestinal inflammation, Enteric bacteria

1 Introduction

Postnatal microbial ontogenesis, which is primarily shaped by key factors such as the gestational age, delivery route and feeding modes has significant consequences on the host immune system early in life as well as adulthood [1, 2]. In the context of prematurity, early microbial colonization of the immature intestine is often associated with the development of a devastating pathology referred to as necrotizing enterocolitis (NEC) [3]. NEC, a deadly disease most commonly affecting very low birth weight (< 1500 g) premature infants, is typically characterized by a plethora of clinical manifestations including feeding intolerance, abdominal distention, excessive gut inflammation and massive destruction of the epithelial barrier [4]. Although the etiology and the pathogenesis of NEC is not well understood, evidence suggests that formula feeding increases the risk of NEC and breast milk plays a protective role [5–7]. We have previously shown that the protection conferred by breast milk is mediated by different components of the milk including epidermal growth factor and the human milk

Belgacem Mihi, Wyatt E. Lanik, and Qingqing Gong contributed equally to this work.

Wendy E. Walker (ed.), *Sepsis: Methods and Protocols*, Methods in Molecular Biology, vol. 2321,
https://doi.org/10.1007/978-1-0716-1488-4_9,

oligosaccharide 2′-fucosyllactose [6, 7]. In addition to the type of enteral feeds received, it is widely accepted that the profile of microbial communities in the intestine of premature babies is a central factor involved in the initiation of NEC. Indeed, several studies have shown that the development of NEC is concomitant with a dysbiosis characterized by a predominance of *Proteobacteria* [2]. It is worth noting that delivery mode, diet and antibiotic administration can profoundly influence the configuration of the intestinal microbial population in premature infants, hence potentially contributing to the development of NEC [2, 8]. In addition to the aforementioned factors, intestinal hypoxic stress seems to play a critical role in triggering NEC as intestinal ischemia has been shown to be a prerequisite to the induction of the disease [9, 10]. Strikingly, it was previously demonstrated that 2-′-fucosyllactose contained in breast milk reduces the severity of NEC through the restoration of intestinal perfusion [7].

Despite the substantial advances made in recent years in our understanding of the pathogenesis of NEC, several obscure facets of the disease need to be elucidated to develop an effective therapeutic strategy. Toward that end, modeling the disease in laboratory animals emerged as the method of choice to unravel the molecular mechanisms involved in NEC pathogenesis. Several investigators have developed different protocols to mimic human disease in multiple animal species [11, 12]. The use of genetically modified mice has proven to be a powerful tool in dissecting different signaling pathways implicated in the development of NEC. The majority of the current mouse models of NEC are based on subjecting neonatal pups to a combination of formula feeds, hypoxia, cold stress, lipopolysaccharide (LPS) and bacterial challenge as elegantly reviewed in [11, 12]. Using a different approach, Zhang et al. developed another model of NEC by combining chemical Paneth cell depletion and *Klebsiella pneumoniae* challenge [13]. Recently, Ginzel and colleagues showed that feeding 3 day old pups with formula containing dextran sodium sulfate results in NEC-like disease affecting the small intestine and the colon [14]. Similarly, another study showed that neonatal mice subjected to the haptenic agent 2,4,6-trinitrobenzene sulfonic acid (TNBS) display intestinal lesions comparable with those observed in humans with NEC [15, 16]. Interestingly, a gene analysis revealed that this model recapitulates the gene signature of the disease in humans [15].

In the current chapter, we will describe a mouse model of NEC based on the use of hypoxic stress and administration of infant formula supplemented with LPS—the ligand for the innate immune receptor toll-like receptor 4 as well as enteric bacteria obtained from a patient with the most severe form of NEC, NEC totalis [17].

2 Materials

1. All animal experiments were performed in accordance with the institutional regulations with Washington University Institutional Animal Care and Use Committee (IACUC). C57BL/6 mice were bred and maintained in a Specific-Pathogen-Free animal facility at Washington University School of Medicine in St. Louis.
2. Isolette Air-Shields Vickers Model C100-200-2 Series 02 Infant Incubator set at 37 °C.
3. 95% nitrogen, 5% oxygen compressed gas tank with a brass single-stage gas regulator connected to hypoxia chamber (Billups-Rothenberg Inc.) and oxygen analyzer (Maxtech Handi+).
4. Frozen glycerol stock of enteric bacteria obtained from an infant with severe surgical NEC (*see* **Notes 1–3**) with approval from the University of Pittsburgh Institutional Review Board.
5. 1.9 French Single-Lumen Silicon Peripherally Inserted Central Catheter (Utah Medical Products).
6. NEC formula (prepare a 2:1 ratio of Abbott Nutrition, Similac Advance OptiGRO with Iron infant formula: Esbilac Puppy Milk replacer, *see* **Note 4**).
7. Autoclaved Luria-Bertani (LB) broth culture medium.
8. Lipopolysaccharide (LPS): Reconstitute LPS from *Escherichia coli* O127:B8 at 10 mg/mL with sterile Dulbecco's phosphate buffered saline without $CaCl_2$ and $MgCl_2$ (PBS). Store at −20 °C.
9. Orbital shaker set at 150 r.p.m. and 37 °C.
10. 15 mL culture tubes with dual position cap.
11. 50 mL conicals.
12. 1.7 mL microcentrifuge tubes.
13. Spectrophotometer with 1 cm cuvette.
14. Semimicro 1.7 mL cuvettes.
15. 1 mL syringes.
16. Polystyrene 100 mm × 15 mm petri dish.
17. 10% neutral buffered formalin.
18. RNAlater™ Solution (Invitrogen).
19. Dissection tools: Scissors, forceps, and dissecting pins.
20. Reusable 24-gauge gavage needle.
21. Macrosette processing/embedding cassettes.
22. Ethanol, isopropanol and chloroform.

23. Steel beads and TissueLyser LT (Qiagen).
24. TRIzol™ Reagent (Invitrogen).
25. Glycogen.
26. QuantiTec Reverse Transcription Kit (Qiagen).
27. IQ SYBR Green Supermix (Bio-Rad).
28. Plate Sealing Film.
29. Real-time PCR detection system.
30. Pedestal spectrophotometer (capable of measuring the concentration of nucleic acids).

3 Methods

3.1 Induction of Necrotizing Enterocolitis in Neonatal Mice

1. The day before the start of the model, take a frozen cryovial of enteric bacteria and gently scrape a small amount of the bacterial glycerol stock with a sterile pipette tip (*see* **Notes 1** and **2**).
2. Transfer the bacteria to a culture tube containing 2 mL of LB and grow the bacteria in an orbital shaker overnight (*see* **Note 3**).
3. The following day at 7 am, transfer 50 μL from the overnight bacterial culture to two culture flasks each containing 20 mL of LB.
4. Culture the inoculum for a minimum of 2 h in a shaking incubator set at 37 °C and 150 r.p.m.
5. In the animal facility, randomly assign 4-day-old C57BL/6 pups into two experimental groups. In the first group, the neonatal mice are kept with their mothers throughout the entire procedure (negative controls), while in the second group, the pups are separated from the dams, weighed, placed in a new cage and maintained in the infant incubator at 37 °C.
6. Measure the optical density of the 2 h culture at 600 nm using a spectrophotometer with a 1 cm cuvette. The OD_{600nm} value should be around 0.6 (corresponding to the exponential phase of bacterial growth).
7. Once appropriate OD_{600nm} is reached, transfer the bacteria to 50 mL conicals, centrifuge for 10 min at 3000 × *g* and discard the supernatant.
8. Resuspend the bacterial pellets in 1 mL PBS.
9. Inoculate 2 mL of bacteria into 20 mL of NEC formula.
10. Based on the average weight of the pups from that morning, add LPS (2.5 μg/g body weight) to the formula after vortexing the LPS for at least 5 min.

A B

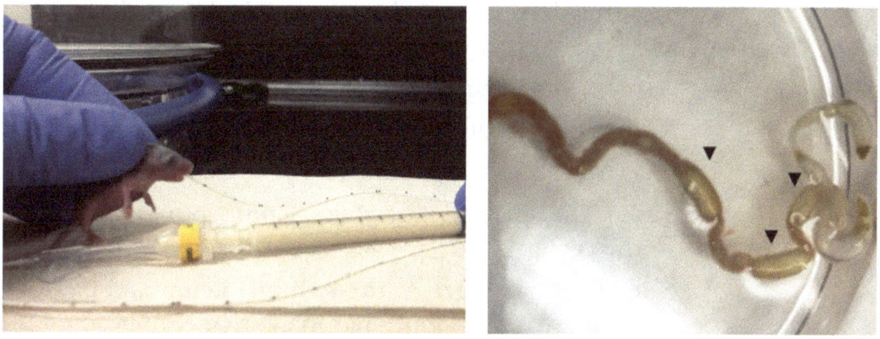

Fig. 1 (**a**) Photograph of a 4-day-old mouse pup during an oral gavage feeding with a PICC line. (**b**) Photograph demonstrating the small intestine and colon of a pup subjected to the NEC protocol. A significant accumulation of gas is observed in the distal part of the gastrointestinal tract (arrowheads) associated with a thin and fragile bowel wall

11. At 10 am, load a 1 mL syringe with the NEC formula and connect it to the PICC line in preparation for the oral gavage.
12. To perform the oral gavage, restrain the mouse by gently grasping the loose skin on the back of the neck. Introduce 1.5 cm of the PICC line into the esophagus through the oral cavity (Fig. 1a). Gavage slowly 80 μL of formula into the pup's stomach. After removing the PICC line, make sure that the animal is breathing properly.
13. The feeding procedure is repeated at 1 pm, 4 pm, 7 pm, and 10 pm.
14. After 1 pm and 7 pm feeds, the pups are subjected to intermittent hypoxia. Pups are placed in a hypoxia chamber (95% nitrogen and 5% oxygen) for 10 min (*see* **Note 5**).
15. Return the pups to their cage and place them in the incubator.
16. Between feeds, the mice should be carefully monitored for apnea, cyanosis, and abdominal distension. Animals presenting these symptoms should be immediately euthanized consistent with IACUC policy.
17. At 3 pm, prepare bacterial culture for the next day as described in **steps 1** and **2**.
18. The following morning, repeat **steps 3** and **4** before the 7 am feed.
19. At the 7 am feed on day 2 and 3, start by weighing the mice. Then feed them with 80 μL of formula (prepared the day before).
20. Repeat **steps 6–17**.

3.2 Sacrificing Animals and Harvesting Tissue Samples

1. Early on day 4, after recording the weights, the pups are euthanized in accordance with IACUC policy.
2. Make a longitudinal incision along the midline of the abdominal cavity to uncover the intestines.
3. Carefully locate and remove the entire small intestine.
4. Place the small intestine in a petri dish filled with cold PBS.
5. Cut the terminal part of the ileum (approximately 2.5 cm from the cecum).
6. Cut a small piece of the terminal ileum and place it in RNAlater™ for transcriptional analysis (*see* Subheading 3.3).
7. Flush the remaining part of the terminal ileum with 10% formalin using a 24-gauge gavage needle (*see* **Note 6**).
8. Put tissue samples in histology processing/embedding cassettes and place them in a 10% formalin container.
9. After 24 h of fixing, the formalin is replaced by 70% ethanol.
10. Embed the intestinal specimens in paraffin by being careful to maintain them as flat as possible.
11. Cut tissue sections on a microtome (5 μm thick) and stain them with hematoxylin and eosin for histological assessment.

3.3 RNA Isolation, cDNA Synthesis, and Transcriptional Analysis of the Intestine

1. Add 1 mL of TRIzol™ to each tissue sample.
2. Add one steel bead per tube and homogenize the tissue using the Qiagen TissueLyser (7 min at 50 oscillations/s). Ensure that intestinal tissues have been completely homogenized.
3. Transfer the homogenized tissues to a 1.7 mL microcentrifuge tube and add 200 ul of chloroform to each tube.
4. Invert the tube 10 times and incubate at room temperature for 5 min.
5. Centrifuge the samples for 15 min at 19,350 × *g*.
6. The solution will separate into three layers. Transfer the top aqueous layer to a new 1.7 mL microcentrifuge tube and add successively 500 μL of isopropanol and 3 μL glycogen. Gently invert the tubes 10 times.
7. Allow the samples to incubate for 20 min at −20 °C.
8. Centrifuge the samples for 15 min at 19,350 × *g*.
9. Remove the supernatant and add 0.5 mL of 75% ethanol.
10. Spin the samples for 5 min at 19,350 × *g*.
11. Remove supernatant and allow the pellet to dry for at least 15 min.

12. Add 50 μL of RNase free water, measure the RNA concentration using a pedestal spectrophotometer and store at −80 °C.
13. Use QuantiTec Reverse Transcription Kit to synthesize the cDNA following the manufacturer's instructions.
14. Real-time PCR (qPCR) is performed using IQ SYBR Green Supermix according to the manufacturer's instructions. The primer sets used in the reaction mix are respectively.

    Il1b Forward primer: AGTGTGGATCCCAAGCAATACCC.

    Il1b Reverse primer: TGTCCTGACCACTGTTGTTTCCCA.

    Cxcl2 Forward primer: TCCAGAGCTTGAGTGTGACG.

    Cxcl2 Reverse primer: CTTCCGTTGAGGGACAGCAG.

    RegIIIg Forward primer: GTACCCTGTCAAGAGCCTCA.

    RegIIIg Reverse primer: TGTGGGGAGAATGTTCCCT.

    RPLO Forward primer: GGCGACCTGGAAGTCCAACT.

    RPLO Reverse primer: CCATCAGCACCACAGCCTTC.

    The qPCR reaction is carried out using a real-time PCR detection system under the following conditions: an initial denaturation step at 95 °C for 3 min, followed by 40 cycles of denaturation at 95 °C for 10 s, primer annealing at 55 °C for 10 s, and primer extension at 72 °C for 30 s. A melting curve can be added to the amplification protocol in order to ensure the absence of primer dimer formation.

3.4 Assessment of NEC-Like Pathologic Intestinal Injury

The protocol described above induces a NEC-like intestinal injury and can result in approximately 30–50% mortality in C57BL/6 Mice (*see* **Notes 7** and **8**). At autopsy, the pups subjected to the NEC model show a large accumulation of gas in the distal part of the gastrointestinal tract (*see* Fig. 1b). The presence of gas is usually associated with a thin and transparent intestinal wall sometimes resulting in an intestinal perforation. Just as in patients with NEC, microscopically as shown in Fig. 2a, the terminal part of the ileum displays mild to complete destruction of the intestinal mucosa. Depending on the disease severity, the ileum is scattered with patchy lesions ranging from a mild destruction of the apical part of the villi to significant transmural necrosis with a complete loss of the intestinal architecture. These histological changes are associated with a significant upregulation of inflammatory factors and antimicrobial peptides such as *Il1b*, *Cxcl2* and *Reg3g* when compared to dam fed controls (Fig. 2b). These histological and transcriptional patterns are also observed in the intestine of human infants with NEC [18, 19].

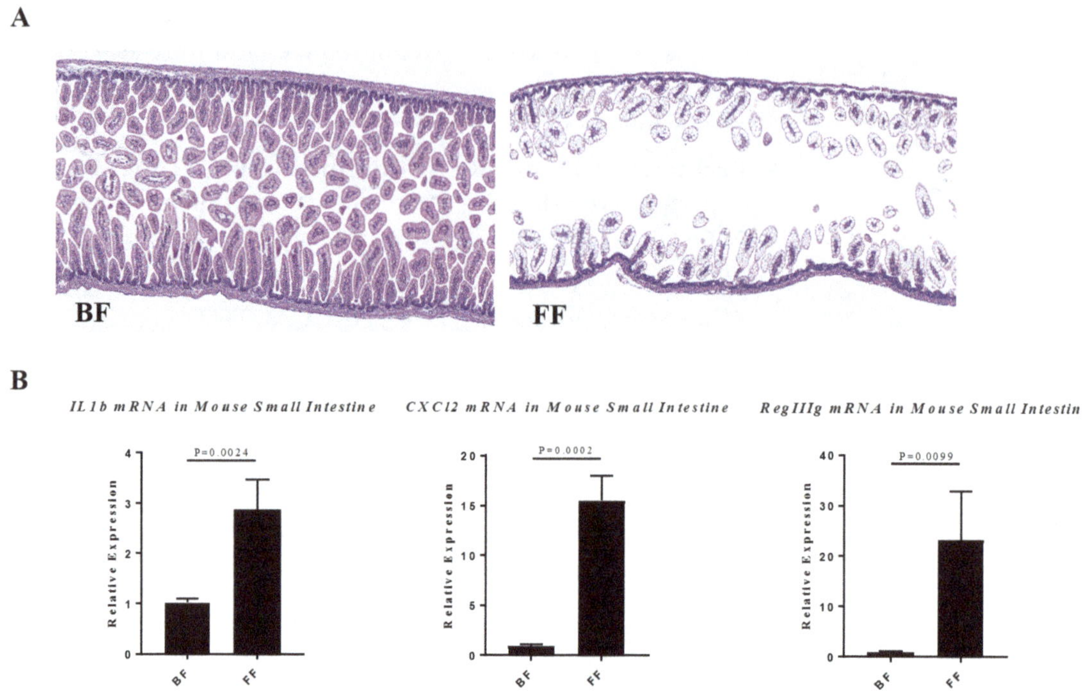

Fig. 2 (**a**) Representative histology of the terminal ileum from a neonatal mouse that was either dam fed (breast fed, BF) or subjected to the NEC protocol (formula fed, FF). (**b**) mRNA gene expression of *Il1b*, *Cxcl2*, and *Reg3g* in the ileum of breast-fed (BF) pups compared to NEC (FF). Statistical significance was computed by the two-tailed Student's t-test

4 Notes

1. The enteric bacterial stock was generated by overnight culture of resected intestinal tissue from an infant with NEC totalis. The overnight culture was centrifuged, resuspended in 50% glycerol and stored at −80 °C.
2. Avoiding excessive freezing–thawing of the bacterial stock is important.
3. A control tube with 3 mL of sterile LB should be cultured overnight in parallel to ensure the absence of bacterial contamination.
4. Do not use old stocks of puppy milk and infant formula.
5. If the pups present with signs of apnea while in the hypoxia chamber, gently shake the hypoxia chamber to stimulate their breathing reflexes. If the apnea persists after the hypoxia treatment, stimulate the breathing reflex by squeezing the terminal part of the pup's tail.

6. Animals with NEC-like intestinal disease have very fragile intestines. Therefore, the intestine has to be slowly flushed in order to avoid causing further damage to the tissue.
7. It is essential to check the integrity of the esophagus and the absence of formula in the lungs during the autopsy. Animals with a ruptured esophagus and milk in the respiratory tract should be excluded from the study and considered as technical deaths.
8. The mortality of this protocol can be higher than 50% in some susceptible mouse strains when subjected to the NEC protocol. A lower LPS concentration or bacterial challenge can reduce the severity of the disease.

Acknowledgments

MG is supported by grants K08DK101608, R03DK111473, and R01DK118568 from the National Institutes of Health, March of Dimes Foundation Grant No. 5-FY17-79, the Children's Discovery Institute of Washington University and St. Louis Children's Hospital, and the Department of Pediatrics at Washington University School of Medicine, St. Louis. Belgacem Mihi, Wyatt E. Lanik, and Qingqing Gong contributed equally to this work.

References

1. Tanaka M, Nakayama J (2017) Development of the gut microbiota in infancy and its impact on health in later life. Allergol Int 66:515–522
2. Warner BB, Tarr PI (2016) Necrotizing enterocolitis and preterm infant gut bacteria. Semin Fetal Neonatal Med 21:394–399
3. Mihi B, Good M (2019) Impact of toll-like receptor 4 signaling in necrotizing enterocolitis: the state of the science. Clin Perinatol 46:145–157
4. Neu J, Walker WA (2011) Necrotizing enterocolitis. N Engl J Med 364:255–264
5. Good M, Sodhi CP, Hackam DJ (2014) Evidence-based feeding strategies before and after the development of necrotizing enterocolitis. Expert Rev Clin Immunol 10:875–884
6. Good M, Sodhi CP, Egan CE et al (2015) Breast milk protects against the development of necrotizing enterocolitis through inhibition of toll-like receptor 4 in the intestinal epithelium via activation of the epidermal growth factor receptor. Mucosal Immunol 8:1166–1179
7. Good M, Sodhi CP, Yamaguchi Y et al (2016) The human milk oligosaccharide 2′-fucosyllactose attenuates the severity of experimental necrotising enterocolitis by enhancing mesenteric perfusion in the neonatal intestine. Br J Nutr 116:1175–1187
8. Denning NL, Prince JM (2018) Neonatal intestinal dysbiosis in necrotizing enterocolitis. Mol Med 24:4
9. Watkins DJ, Besner GE (2013) The role of the intestinal microcirculation in necrotizing enterocolitis. Semin Pediatr Surg 22:83–87
10. Yazji I, Sodhi CP, Lee EK et al (2013) Endothelial TLR4 activation impairs intestinal microcirculatory perfusion in necrotizing enterocolitis via eNOS-NO-nitrite signaling. Proc Natl Acad Sci U S A 110(23):9451–9456
11. Sodhi C, Richardson W, Gribar S et al (2008) The development of animal models for the study of necrotizing enterocolitis. Dis Model Mech 1:94–98
12. Sulistyo A, Rahman A, Biouss G et al (2018) Animal models of necrotizing enterocolitis: review of the literature and state of the art. Innov Surg Sci 3:87–92
13. Zhang C, Sherman MP, Prince LS et al (2012) Paneth cell ablation in the presence of

Klebsiella pneumoniae induces necrotizing enterocolitis (NEC)-like injury in the small intestine of immature mice. Dis Model Mech 5:522–532

14. Ginzel M, Feng X, Kuebler JF et al (2017) Dextran sodium sulfate (DSS) induces necrotizing enterocolitis-like lesions in neonatal mice. PLoS One 12:e0182732
15. MohanKumar K, Namachivayam K, Cheng F et al (2017) Trinitrobenzene sulfonic acid-induced intestinal injury in neonatal mice activates transcriptional networks similar to those seen in human necrotizing enterocolitis. Pediatr Res 81:99–112
16. MohanKumar K, Kaza N, Jagadeeswaran R et al (2012) Gut mucosal injury in neonates is marked by macrophage infiltration in contrast to pleomorphic infiltrates in adult: evidence from an animal model. Am J Physiol Gastrointest Liver Physiol 303:G93–G102
17. Good M, Sodhi CP, Ozolek JA et al (2014) Lactobacillus rhamnosus HN001 decreases the severity of necrotizing enterocolitis in neonatal mice and preterm piglets: evidence in mice for a role of TLR9. Am J Physiol Gastrointest Liver Physiol 306:G1021–G1032
18. Ballance WA, Dahms BB, Shenker N et al (1990) Pathology of neonatal necrotizing enterocolitis: a ten-year experience. J Pediatr 117:S6–S13
19. MohanKumar K, Namachivayam K, Ho TT et al (2017) Cytokines and growth factors in the developing intestine and during necrotizing enterocolitis. Semin Perinatol 41:52–60

Chapter 10

A Murine Model of Full-Thickness Scald Burn Injury with Subsequent Wound and Systemic Bacterial Infection

Antonio Hernandez, Naeem K. Patil, and Julia K. Bohannon

Abstract

Infection is the leading cause of death and prolonged hospitalization in severely burned patients that survive the acute phase of injury. Here we describe a murine model of severe burn injury followed by subsequent postburn infection, both local and systemic, that leads to sepsis. A detailed description of the full-thickness scald burn procedure is provided, followed by description of infection with two common burn-associated nosocomial pathogens, *Pseudomonas aeruginosa* and *Staphylococcus aureus*.

Key words Burn, Sepsis, *Pseudomonas aeruginosa*, *Staphylococcus aureus*, Nosocomial infection

1 Introduction

Despite the use of both topical and systemic antibiotics, infection remains the most common cause of prolonged hospitalization and death in burn patients who survive the initial phase of systemic shock and resuscitation [1]. Factors that predispose burned patients to infections are complex. Numerous reports have shown that severe burn injury induces immune dysfunction leading to increased host susceptibility for severe infections [2–4]. Destruction of the skin barrier and the emergence of antibiotic resistant pathogens are additional factors [5, 6]. In 2013, the Centers for Disease Control (CDC) listed two of the most common drug resistant pathogens known to threaten burn patients, *Pseudomonas aeruginosa* and *Staphylococcus aureus*, as "Serious Threats" requiring increased monitoring and prevention strategies [7–9]. Effective clinically relevant animal models to mimic the scenario of postburn infection and sepsis are required to better understand immune dysfunction following severe burn injury and to aid in the development of therapies intended to enhance resistance to infection.

Wendy E. Walker (ed.), *Sepsis: Methods and Protocols*, Methods in Molecular Biology, vol. 2321,
https://doi.org/10.1007/978-1-0716-1488-4_10,

Here, we describe our well-established murine burn wound model [10–13]. Briefly, under isoflurane anesthesia, the shaved dorsum of mice is exposed to 98 °C water for 10 s followed by fluid resuscitation to induce a burn injury corresponding to ~30–40% of the total body surface area (TBSA). For the induction of burn wound infection, 1×10^6–1×10^8 colony forming units (CFU) of *P. aeruginosa* is topically applied to the burn wound on day 4 postburn. This model results in reproducible dissemination of infection throughout the burn wound and into the systemic circulation and organs within 3–6 days after wound inoculation for BALB/c mice [10, 11, 14]. *P. aeruginosa* has a high rate of antibiotic resistance and is highly prevalent as a nosocomial pathogen, particularly in burn patients [15, 16]. The CDC defines *P. aeruginosa* as one of the most common causes of nosocomial infections and of particular concern due to its extraordinary resistance to antimicrobials. In a cohort of pediatric burn patients at the Shriners Hospital in Galveston, Williams et al. showed that 47% of mortality in burn patients was due to sepsis and that 73% of septic patients were infected with antibiotic resistant organisms. *Pseudomonas aeruginosa* was responsible for 64% of deaths caused by antibiotic resistant bacteria in their cohort [17].

S. aureus is one of the primary organisms responsible for gram-positive infections in burn patients [18]. Like *P. aeruginosa*, *S. aureus* is identified by the CDC as a high threat organism due to its high incidence of antibiotic resistance [7–9]. Our previous studies show that intravenous *S. aureus* challenge induces reproducible morbidity and mortality after intravenous administration following burn injury [19]. In the burn injury model described in this chapter, the strain of *S. aureus* used does not lead to systemic infection and sepsis when applied topically at the wound site, but consistently induces lethal systemic infection when administered intravenously. *S. aureus* has been reported as the most common organism isolated from blood cultures of burn patients with sepsis [20].

Thus, we describe a highly reproducible model of full-thickness scald burn injury in mice. We further describe two clinically relevant postburn infection models. A variety of other multidrug-resistant organisms are known to contribute to morbidity and mortality following severe burn injury, including *Acinetobacter* spp. and fungal pathogens including *Candida* spp. and *Aspergillus* spp. [20]. While these pathogens are not covered in this chapter, model development in the context of burn injury for these pathogens is warranted.

2 Materials

2.1 Burn Injury

1. Water bath.
2. Electric razor with a size 40 blade.
3. Platform for burn injury:

   We use a hard-plastic custom-made platform, which can withstand boiling water temperatures (Fig. 1). The size of the platform should be individualized for particular mouse size and percentage of burn injury desired. The image shown depicts a prototype platform used for a 10–12 weeks old mouse, weighing approximately 25 g, to induce a ~30–40% TBSA burn injury. Surface area was calculated by measurement of an excised mouse pelt, and was confirmed with a calculation that is based on weight, using a published constant for mice as described [21].
4. 12 mL syringes.
5. Normal saline (NS) (sterile).
6. Ringer's lactate (LR) (sterile).
7. Needles: Gauge 26 and Gauge 23.
8. 2 × 2 in. alcohol swabs
9. Anesthesia: Isoflurane.
10. Analgesia: Buprenorphine.
11. Anesthesia chamber and machine with plastic tubing.
12. Timer.
13. Paper towels.
14. Sterile cages.

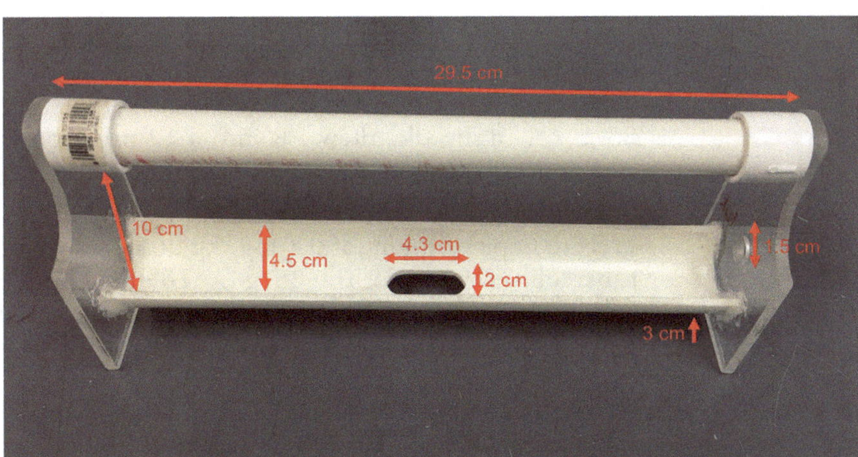

Fig. 1 Prototype of a platform for inducing scald burn injury in a mouse model. All dimensions are in centimeters

2.2 Infection

1. *Pseudomonas aeruginosa* (Schroeter) Migula (American Type Culture and Collection, ATCC 19660).
2. *Staphylococcus aureus* subsp. *aureus* Rosenbach (ATCC 25923).
3. Tuberculin syringe, 0.5 mL, with attached 27G needle.
4. Tryptic soy agar plates (prepared according to the manufacturer's protocol: Sigma Aldrich #22091).
5. Trypticase Soy Broth.
6. Petri dishes.
7. T or L shaped spreader.
8. Conical polystyrene centrifuge tubes (15 and 50 mL).
9. Laboratory centrifuge.
10. Orbital shaker (either Incubated shaker, or shaker placed inside incubator).
11. Glass or plastic flask (125 mL).
12. Pipettes: (100 μL, 200 μL, and 1 mL) and respective sterile tips.
13. Bacterial incubator.

3 Methods

3.1 Material Preparation

1. Spinal cord protection: Requires 1 mL of 0.9% normal saline (NS) per mouse. Take a 2 × 2-in. alcohol swab and cleanse the injection port of the sterile NS bag, then using a sterile 21 g × 1.5-in. needle, draw saline into a 12 mL syringe. Prepare appropriate volume of NS and number of syringes based on the number of mice.
2. Intraperitoneal (IP) fluid resuscitation: Requires 2 mL of Ringer's lactate (LR) solution per mouse. Take an alcohol swab and wipe the injection port of the sterile LR bag, then using a sterile 23 g × 1-in. needle, draw LR into a 12 mL syringe. Prepare appropriate volume of LR and number of syringes based on the number of mice.
3. Prepare buprenorphine at 0.1 mg/kg dose, drawn into a 1 mL tuberculin syringe with a 26 g × 10 mm needle. Prepare appropriate volume of buprenorphine and number of syringes based on the number of mice and their weight.

3.2 Anesthesia and Analgesia Preparation

1. Administer buprenorphine at 0.1 mg/kg, 30 min prior to initiation of burn injury, by subcutaneous injection into the nuchal region.

2. Transfer the mouse to the anesthesia chamber.
3. Using fresh gas flow rate of 1 L/min of oxygen through the anesthesia machine, initiate anesthesia using 2.5% isoflurane.

3.3 Burn Wound Injury

1. Remove a single mouse from the anesthesia chamber and place it on its abdomen, with the mouse's nose in the ETT adaptor to maintain 2.5% isoflurane anesthesia.
2. Shear the fur from the mouse's dorsum.
3. Inject 1 mL of NS subcutaneously along the full length of the vertebral column (base of nuchal region to base of caudal region).
4. Place the mouse on the burn fenestrated platform, in a supine position, exposing the area that will be burned (Fig. 2).
5. Immediately place the mouse's nose into the tubing adaptor attached to the platform, to maintain 2.5% isoflurane anesthesia at 1 L/min oxygen flow directly from the anesthesia machine for the duration of the burn injury induction.
6. Ensure that only the sheared skin should be protruding through the platform fenestration.
7. Verify that both ears are above the fenestration to avoid unnecessary injury to the mouse ear. Confirm a tight fit through the fenestration to avoid water surfacing from the edge of the fenestration.
8. Confirm water bath is >97.0 °C but <99.0 °C.
9. Burn the mouse by submerging the fenestrated platform for exactly 10 s using a timer.

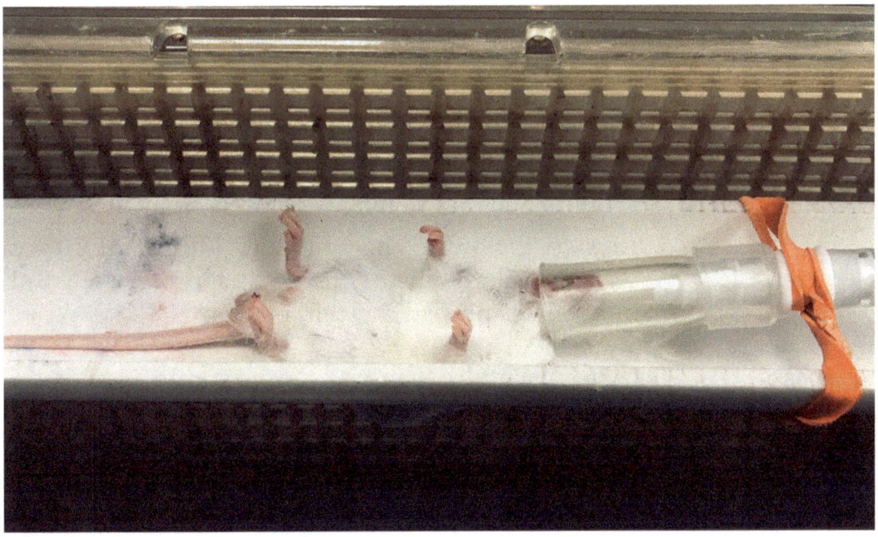

Fig. 2 Burn platform with mouse and anesthesia setup

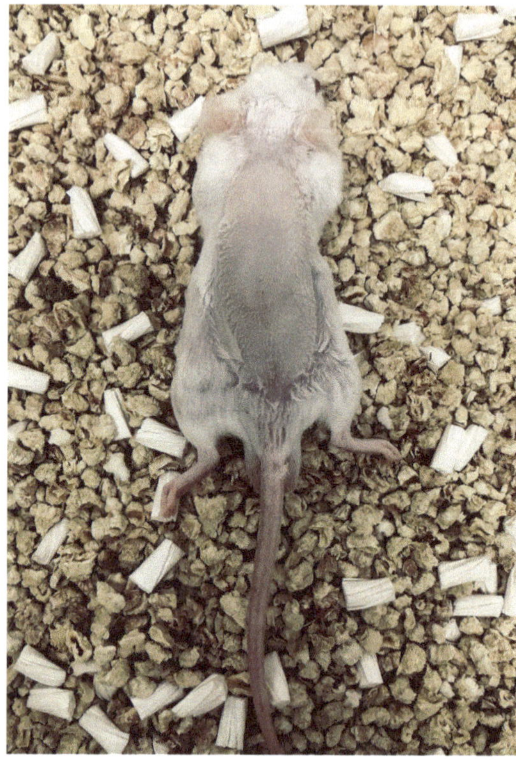

Fig. 3 Burn area corresponding to ~30–40% TBSA for a 10-week-old male BALB/c, 23 g mouse

10. Immediately remove the mouse from the water bath and blot the burned area with paper towels to avoid continued injury.
11. Wipe the mouse's abdomen with an alcohol swab.
12. Inject 2 mL of LR into the peritoneum for fluid resuscitation, based on the Parkland formula for burns [22].
13. Place the mouse into an isolated cage, in the prone position, to avoid contamination of the burn wound (*see* **Note 1**).
14. *See* Fig. 3 for corresponding 30–40% TBSA scald burn.
15. Monitor the mouse until it recovers from anesthesia.
16. Reassess the burned mice and provide 0.1 mL (0.1 mg/kg) of buprenorphine subcutaneously at approximately 8–12 h post-burn and as needed thereafter.

3.4 Pseudomonas aeruginosa *and* Staphylococcus aureus *Culture Preparation*

1. Bacterial culture setup: 3 days prior to inoculation, grow a stock culture of *P. aeruginosa* or *S. aureus* from stock vials stored at −80 °C. Thaw vial and add to a flask with 30 mL of refrigerated TSA broth. Leave lid loose on flask and place in the shaker at 37 °C. Turn shaker on at 200 rpm for exactly 22 h.

2. Exactly 22 h after placing the stock culture in the shaker, pour culture into a 50 mL conical tube.
3. Centrifuge at 5000 × *g* for 15 min, at 4 °C.
4. Discard supernatant and disrupt the bacterial pellet by gently tapping the tube.
5. Wash pellet with 30 mL of sterile NS and vortex well to resuspend.
6. Centrifuge as described above.
7. Pour off supernatant, disrupt the bacterial pellet as described above.
8. Add 2.5 mL sterile NS and vortex well. This is the stock culture going forward.
9. Perform serial dilutions in sterile NS for plating:

   Set up dilution tubes:

   10^{-2}: 9.9 mL NS + 0.1 mL stock (2.5 mL) culture, vortex

   10^{-4}: 9.9 mL NS + 0.1 mL 10^{-2} dilution, vortex

   10^{-6}: 9.9 mL NS + 0.1 mL 10^{-4} dilution, vortex

   10^{-8}: 9.9 mL NS + 0.1 mL 10^{-6} dilution, vortex.
10. Plate 0.1 mL of the 10^{-8} dilution on two plates (prepared according to manufacturer's protocol), spread thoroughly and place in incubator.
11. After 10 min, invert plates, incubate overnight.
12. On the following day, count colonies on the two plates, and calculate the average.
13. Determine colony forming units of stock bacterial culture:

$$\text{Average colonies} \times \text{dilution factor}(10^8) \times 10(\text{plating dilution}) = \text{stock CFU}$$

Dilute appropriately for working inoculation concentration (*see* **Note 2**).

3.5 Pseudomonas aeruginosa *Burn Wound Infection*

1. Prepare 1 × 10^6 to 1 × 10^8 CFU in 0.05 mL of sterile NS for wound inoculation. Dilute stock culture prepared as described above (*see* **Notes 3** and **4**).
2. Four days post–burn injury (*see* **Note 5**), sedate the animal under 2.5% isoflurane anesthesia.
3. Apply the inoculum at the center of the burn wound using a pipet and evenly and gently distribute throughout the area with pipet tip, while avoiding ~3 mm wound margins (this margin area is likely to not be full-thickness burn).
4. Place the animal prone to maintain the inoculum intact until the mouse emerges from anesthesia.

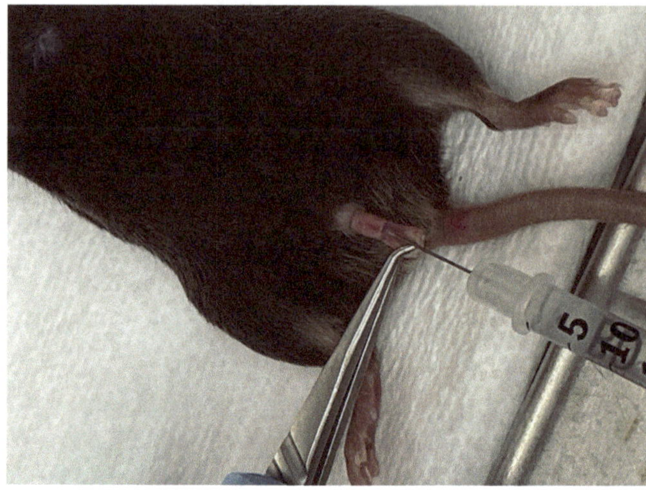

Fig. 4 Penile vein injection. Mouse is placed under continuous isoflurane anesthesia, using forceps to gently retract the penis, injection into the dorsal vein of the penis is performed with a 27G needled affixed to a 0.5 mL tuberculin syringe

5. Place dampened chow in the cage to facilitate access to nutrition, as the infected wound limits mobility of the mice over time as the infection spreads.
6. Examine the mice until they have recovered from anesthesia.

3.6 Staphylococcus aureus *Intravenous Systemic Infection after Burn*

1. Prepare 1×10^8 CFU in 0.2 mL of NS for intravenous systemic infection. Dilute stock culture as described above.
2. Four days post–burn injury (*see* **Note 5**), vortex the bacterial suspension, and using a tuberculin syringe, aspirate 0.2 mL of the *S. aureus* dilution and inject into the penile vein under 2.5% isoflurane anesthesia (Fig. 4) [23]. Alternatively, tail vein injection may be appropriate if using female mice.
3. Place mice into a cage and examine until recovery from anesthesia.

4 Notes

1. Post–burn injury, mice MUST be housed separately to avoid chewing to the burn wound by another mouse.
2. Bacterial concentration can also be calculated by obtaining an optical density reading using a spectrophotometer.
3. For the *P. aeruginosa* burn wound infection, the concentration of the bacterial inoculum varies depending on the size of the mice, strain of mice, size of the burn injury, and lethality of the

organism strain. Due to this variability, optimization of an appropriate dose may be required. This protocol is optimized for ~25–30 g, 10–12-week-old male BALB/c mice.

4. Significantly higher pathogen loads are required to cause illness in the immune-competent nonburned animals. Particularly with the burn wound infection model, it is not possible to induce a wound infection in the absence of burn injury.

5. Our protocol is optimized for induction of *P. aeruginosa* and *S. aureus* infection on day 4 post–burn injury. This falls within the common clinical scenario for incidence of primary wound infections and bloodstream infections, which commonly occur within the first week, yet beyond the first 24–48 h following burn injury [18]. However, the infection timing can be altered to an earlier or later time point depending on experimental query. If altering the day of infection, the dose may need to be adjusted and optimization will be necessary.

Acknowledgments

This work was supported by NIH grant R01GM121711 to Julia K. Bohannon, NIH grant K08 GM123345 to Antonio Hernandez, and Shock Society Faculty Research Award and Vanderbilt Faculty Research Scholars Award to Naeem K. Patil.

References

1. Polk HC (1979) Supportive therapy in burn care. Consensus summary on infection. J Trauma 19:894–896
2. Murphey ED, Sherwood ER, Toliver-Kinsky T, Herndon D (2007) The immunological response and strategies for intervention. In: Total burn care. Saunders Elsevier, Philadelphia, PA, pp 310–324
3. Finnerty CC, Herndon DN, Chinkes DL et al (2007) Serum cytokine differences in severely burned children with and without sepsis. Shock 27:4–9
4. Fitzwater J, Purdue GF, Hunt JL et al (2003) The risk factors and time course of sepsis and organ dysfunction after burn trauma. J Trauma 54:959–966
5. Geyik MF, Aldemir M, Hosoglu S et al (2003) Epidemiology of burn unit infections in children. Am J Infect Control 31:342–346
6. Bang RL, Sharma PN, Sanyal SC et al (2002) Septicaemia after burn injury: a comparative study. Burns 28:746–751
7. Solomon SL, Oliver KB (2014) Antibiotic resistance threats in the United States: stepping back from the brink. Am Fam Physician 89:938–941
8. Allen J (2013) The threat of antibiotic resistance. Geriatr Nurs 34:519–520
9. Hampton T (2013) Report reveals scope of US antibiotic resistance threat. JAMA 310:1661–1663
10. Bohannon JK, Luan L, Hernandez A et al (2016) Role of G-CSF in monophosphoryl lipid A-mediated augmentation of neutrophil functions after burn injury. J Leukoc Biol 99:629–640
11. Bohannon J, Cui W, Cox R et al (2008) Prophylactic treatment with fms-like tyrosine kinase-3 ligand after burn injury enhances global immune responses to infection. J Immunol 180:3038–3048
12. Bohannon J, Cui W, Sherwood E et al (2010) Dendritic cell modification of neutrophil responses to infection after burn injury. J Immunol 185:2847–2853

13. Bohannon JK, Luan L, Hernandez A et al (2015) Role of G-CSF in monophosphoryl lipid A-mediated augmentation of neutrophil functions after burn injury. J Leukoc Biol 99 (4):629–640
14. Toliver-Kinsky TE, Cui W, Murphey ED et al (2005) Enhancement of dendritic cell production by fms-like tyrosine kinase-3 ligand increases the resistance of mice to a burn wound infection. J Immunol 174:404–410
15. Sun HY, Fujitani S, Quintiliani R et al (2011) Pneumonia due to Pseudomonas aeruginosa: part II: antimicrobial resistance, pharmacodynamic concepts, and antibiotic therapy. Chest 139:1172–1185
16. Williams FN, Herndon DN, Hawkins HK et al (2009) The leading causes of death after burn injury in a single pediatric burn center. Crit Care 13:R183
17. Williams FN, Herndon DN, Hawkins HK et al (2009) The leading causes of death after burn injury in a single pediatric burn center. Crit Care 13:R183
18. Church D, Elsayed S, Reid O et al (2006) Burn wound infections. Clin Microbiol Rev 19:403–434
19. Patil NK, Luan L, Bohannon JK et al (2018) Frontline science: anti-PD-L1 protects against infection with common bacterial pathogens after burn injury. J Leukoc Biol 103:23–33
20. Branski LK, Al-Mousawi A, Rivero H et al (2009) Emerging infections in burns. Surg Infect 10:389–397
21. Traber DL, Barrow RE, Herndon D (2001) Animal models of burn injury. In: Souba W, Wilmore D (eds) Surgical research. Academic Press, San Diego, pp 367–377
22. Mehta M, Tudor GJ (2020) Parkland Formula. StatPearls, Treasure Island, FL
23. Nebendahl K, Hauff P (2011) Drug administration. In: Kiessling F, Pichler BJ (eds) Small Animal Imaging: Basics and Practical Guide. Springer, Berlin, pp 93–116

Chapter 11

Mouse Intensive Care Unit (MICU)

Tamara Merz, Sandra Kress, Michael Gröger, Peter Radermacher, and Oscar McCook

Abstract

The translation of preclinical results into successful clinical therapies remains a challenge in sepsis research. One reason for this lack of translation might be the discrepancy between preclinical models and the clinical reality: nonresuscitated young healthy rodents in contrast to elderly comorbid patients in an intensive care unit. We introduce the mouse intensive care unit (MICU) as a concept to address the lack of resuscitation in preclinical studies as one of the limiting issues in translational research. The MICU reflects standard procedures of the clinical intensive care unit: fluid resuscitation, lung-protective mechanical ventilation, and hemodynamic monitoring and management, all tailored to organ- and function-specific targets. Thus, the MICU gives an experimental animal the intermediate possibility of recovery and survival due to "patient" management, which is not reflected in less complex experimental scenarios, which either result in acute survival or death.

Key words Preclinical animal model, Translational medicine, Intensive care, Sepsis, Circulatory shock, Polytrauma

1 Introduction

1.1 Why Do We Need a MICU?

20 years ago, the late Daniel Traber, who significantly influenced our ICU concept, already pointed out that no critical care physician would accept data from a nonresuscitated septic patient, nor drug efficacy data from ICU studies that did not include resuscitation, blood pressure or heart rate monitoring [1]. Yet recently, an international expert consensus on the Minimum Quality Threshold in Pre-Clinical Sepsis Studies (MQTiPSS) reported that, among the most highly cited studies between 2003 and 2017, over 70% did not provide any fluid resuscitation at all and, of the less than 30% that did provide fluids, only 3% titrated the resuscitation to a specific organ support [2, 3]. In spite of this fact, in 1999 we were reminded that "the pathophysiology and outcome of an unresuscitated, unmonitored, septic patient is certainly different" [1]. The significant discrepancies between preclinical studies and

Wendy E. Walker (ed.), *Sepsis: Methods and Protocols*, Methods in Molecular Biology, vol. 2321,
https://doi.org/10.1007/978-1-0716-1488-4_11,

the clinical reality limit their potential impact and translational value [3], as reflected in many failed sepsis trials [4]. In 2001, Hollenberg et al. underscored the importance of basic resuscitation, showing that the simple administration of fluids and antibiotics in a murine model of polymicrobial sepsis increased 48 h survival from 0% to 46%, which is comparable to the clinical reality [5, 6]. The most common fluid resuscitation used in sepsis is saline (0.9% sodium chloride) which can lead to hyperchloremic metabolic acidosis [7, 8]. Recently balanced electrolyte solutions have been shown to reduce the risk of hyperchloremic metabolic acidosis, acute kidney injury, inflammation, and death, both in preclinical and clinical models [7, 8]. Iso-osmolar balanced crystalloid solutions are preferred, as recommended by the recent international MQTiPSS consensus [3, 7, 8]. In an effort to improve the relevance for translational medicine, there has been a push toward more complex animal modeling, which would attempt to replicate the current clinical practice in the intensive care unit (ICU) as closely as possible [4, 6].

An ICU helps to sustain life during a period of life-threatening organ system dysfunction by monitoring of critically ill patients and providing intensive and specialized medical care and physiologic organ support [9]. For example, basics in septic patient management following the Surviving Sepsis guidelines [10, 11] comprise anesthesia, antibiotics, control of glycemia and temperature, "lung-protective" mechanical ventilation, hemodynamic monitoring with adequate fluid resuscitation as well as vasopressor/inotropic support based on need (target mean arterial pressure $\geq$ 65 mmHg (or $\geq$55 mmHg for mice [12]) [6]. Today, the technological advances allow us to mimic the ICU in mice, that is, set up a "mouse ICU" (MICU), through the use of miniaturized instruments: catheters, vascular flow-probes, and ventilation tubing tailored to small animals [6]. An intravenous catheter, for example, provides an opportunity to administer deep anesthesia, crystalloids, catecholamines, stable isotopes and experimental drugs; an intra-arterial catheter can be used for repetitive blood sampling (for blood gas analysis, essential for "patient management" in the ICU) and to measure heart rate and blood pressure, or even left-ventricular function, if the catheter is supplied with pressure and conductance sensors; the bladder catheter facilitates urine sampling. Flow probes for macro- and microvasculature give us a handle on tissue oxygen availability and blood flow: microvascular perfusion, capillary hemoglobin concentration, and oxygen saturation [13]. This information has proven to be particularly relevant in sepsis experiments, where it has been a matter of debate whether organ failure is related to tissue hypoxia (inadequate oxygen delivery) or the cellular inability to utilize oxygen [14, 15]. Implementing the MICU in a CLP model of septic shock, Merz et al. performed repetitive blood gas analysis and titrated norepinephrine to the MAP ($\geq$55 mmHg) [16]. Thus they not only assured that the animals were septic but also met the acute circulatory failure criteria proposed by the Sepsis-3 guidelines in response to septic shock: lactate $\geq$2 mmol/l and catecholamine requirements to maintain MAP in spite of adequate fluid resuscitation [11, 16]. In addition, they were able to determine that liver injury was indeed due to reduced mitochondrial activity and not due to insufficient tissue oxygenation, which could only be measured by the use of macro- and microvascular flow probes [16].

Lung-protective mechanical ventilation was crucial to the maintenance of proper tissue oxygen supply in the above mentioned experiment [16] and is also a standard life-saving procedure in the ICU [15, 17]; thus, ventilation is an essential component of a translationally relevant experimental setup. In fact, in mechanical ventilation studies, we find a particularly striking example of a preclinical experiment translating into profound changes in clinical practices. In rats undergoing mechanical ventilation (45 cm H_2O peak airway pressure), it was observed that "end-inspiratory distension is the main determinant of VILI," thus leading to the "reduction of tidal volume [in the clinic] and improved survival of ARDS patients" [18, 19]. It is ironic that such a simple physiological modification of clinical practice, first demonstrated in a rodent model [20], improves the outcome of many critically ill patients and has thus had much more impact than many large and expensive drug trials [18, 19].

Another recommendation from the recent MQTiPPS consensus is the addition of comorbidities to our preclinical experiments [2]. This is not a new concept: researchers have called for including older animals [21], genetic diversity, different sexes [22], preexisting disease, two hit insults [23, 24], and chronic exposure protocols [25] in animal modeling [4, 26]. An epidemiological study suggests that adult septic patients with underlying comorbidities have a 3.5-fold increase in mortality in comparison to noncomorbid patients [27].

By implementing the ICU in animal modeling through monitoring, treatment, and organ support, we can reproduce a key factor that is missing from preclinical sepsis models. While most animal studies focus exclusively on mortality as an endpoint, the MICU delivers an intermediate possibility of surviving due to patient management, which represents a significant advance over a simplified scenario of natural acute death versus recovery [6]. Even though these strategies will tremendously increase the procedural complexity, they are crucial in counteracting the current mismatch between preclinical research and real-life patient care [6].

2 Materials

2.1 Drugs and Reagents for Veterinary Care

1. Eye ointment.
2. Anesthetics: ketamine, midazolam, fentanyl.
3. Crystalloids: balanced electrolyte solution.
4. Colloids: hydroxyethyl starch in balanced electrolyte solution.
5. Vasopressors: norepinephrine.

2.2 Surgical Supplies (See Fig. 1c)

1. Gloves.
2. Cotton swabs.
3. Gauze.
4. Suturing material.
5. Syringes.
6. Three-way valves.
7. Catheters: vascular lines and left-ventricular catheter (with pressure and conductance sensor, 1.4 Fr).
8. Ventilation tube and connecting lines.
9. Flow probes.
10. Glass capillaries, magnets, and stoppers for blood gas analysis.

2.3 Instruments (See Fig. 1a, b)

1. Hair clipper.
2. Microsurgical tools: surgical scissors, surgical forceps, needle holder.
3. Dissection microscope.
4. Controlled-rate infusion pumps.
5. Closed-loop temperature control system (rectal temperature probe, heat mat and thermometer with feedback loop).
6. Small animal ventilator and monitor.
7. Heart/blood pressure monitor (measuring amplifier).
8. Blood gas analyzer.
9. Pressure–volume analog signal amplifier.

3 Methods

3.1 Surgical Preparation

1. Prepare the workspace (*see* Fig. 1).
2. Weigh the animal (*see* **Note 1**).
3. Sedate the animal via exposure to sevoflurane in an anesthetic chamber (*see* **Note 2**).

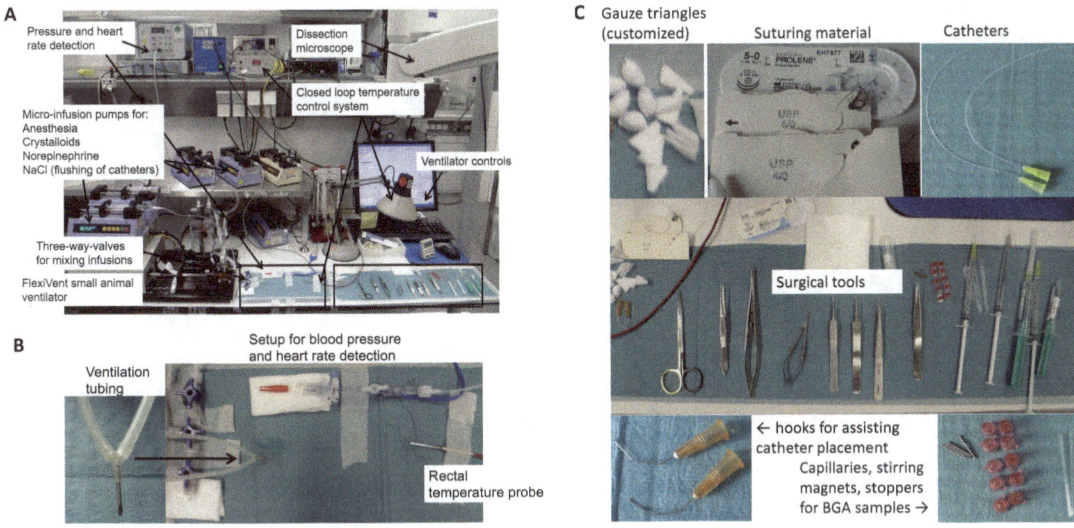

Fig. 1 View of mouse intensive care unit (MICU) prepared for an experiment. (**a**) shows the setup with all the instruments in the mouse ICU. The black box on the left represents the space where the mouse will be placed. Underneath the surgical drape a heat mat is positioned; (**b**) is a higher magnification; (**c**) is a higher magnification of the black box to the right displaying all needed consumables and surgical tools

4. Administer i.p. anesthesia (120 μg/g ketamine, 1.25 μg/g midazolam, 0.25 μg/g fentanyl) (*see* **Note 3**).
5. Clip the hair of all surgical sites (*see* **Note 4**).
6. Administer eye ointment (*see* **Note 5**).
7. Continuously measure the animal's rectal temperature with a temperature probe. Place the animal on a heat mat with a closed-loop temperature control system (*see* **Note 6**).

3.2 Tracheostomy (See Fig. 2)

1. Expose the trachea by performing a longitudinal cut in the skin above the trachea and perform blunt dissection by pulling apart the muscle tissue (*see* **Notes** 7, **8** and Fig. 2a).
2. Pull a suture around the trachea to provide for a loose ligature.
3. Make a slight incision in the trachea and insert an endotracheal tubing (*see* **Note 9**, Fig. 2b, c).
4. Fix the tubing by tightening the ligature (*see* Fig. 2d) and connect the tubing to the ventilator.
5. Perform a recruitment maneuver (*see* **Note 10**, and Subheading 3.8).
6. Begin lung-protective mechanical ventilation (initial settings: $F_iO_2 = 0.21$, respiratory rate 150 breaths/min, tidal volume 4–6 μl/g, inspiratory/expiratory time ratio 1:2, and positive end-expiratory pressure (PEEP) 3 cmH_2O, *see* Subheading 3.8 for details on adjusting the ventilator settings).

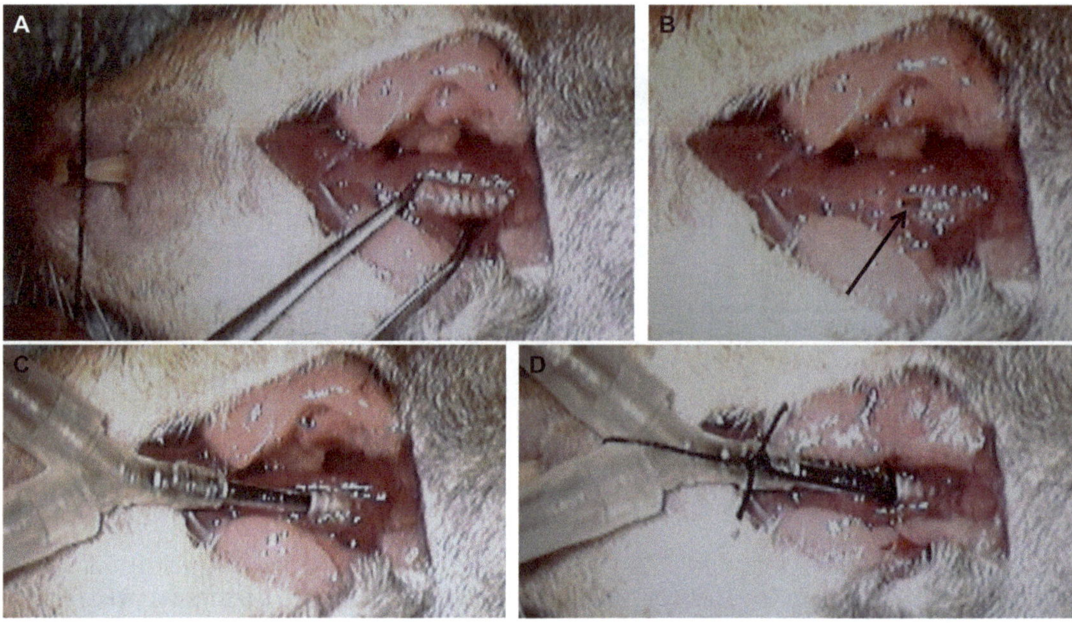

Fig. 2 Endotracheal intubation. (**a**) Exposure of the trachea. (**b**) Incision (arrow). (**c**) Endotracheal intubation. (**d**) Fixation of the tubing with sutures

3.3 Cannulation of Right External Jugular Vein (See Fig. 3)

1. Make an incision in the skin over the right external jugular vein, exposing approximately 7 mm of the vein (*see* Fig. 3a).
2. Prepare two ligations: grab suture with curved microforceps and prick underneath the vein, put an open loop around the vessel (*see* **Note 11** and Fig. 3b).
3. Tie the distal ligation (closer to the animal's head) using a surgeon's knot.
4. Using a slanted incision, open up the vessel between the ligations with microscissors (*see* **Note 12**).
5. Ease a catheter 1–2 cm into the vein (*see* **Note 13**, Fig. 3c–e). Test if blood can be drawn easily. Then tie the second ligation (two surgeon's knots, *see* Fig. 3f). Recheck patency.
6. Start an i.v. infusion of ketamine 100–150 μg per g body weight per hour [μg/(g h)], midazolam 0.2–0.3 μg/(g h), and fentanyl 1–1.25 μg/(g h) (*see* **Note 14**).

3.4 Cannulation of Right Carotid Artery

1. Make an incision in the skin over the right carotid artery. Perform a blunt deep dissection of the artery (*see* **Notes 15** and **16**).
2. Separate the artery from the vagus nerve.
3. Expose approximately 7 mm of the artery (*see* Fig. 4a).
4. Prepare two ligations (similar to jugular vein cannulation, *see* Fig. 4b, c).

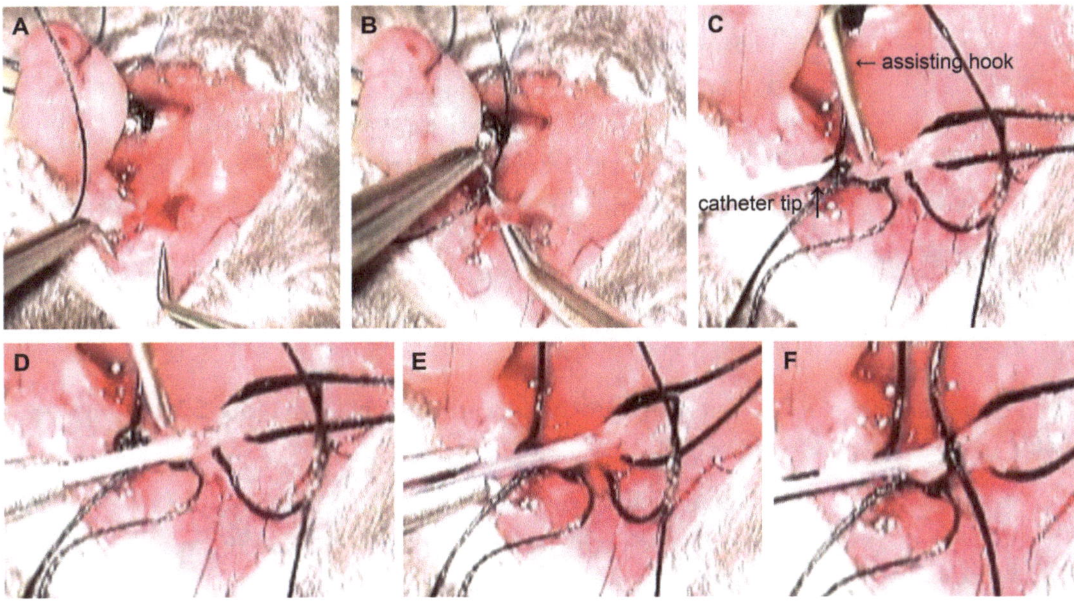

Fig. 3 Placement of catheter in the jugular vein. (**a**) Exposure of the jugular vein. (**b**) Preparation of ligation. (**c**) Incision. (**d**) Introduction of catheter. (**e**) Catheter in place. (**f**) Catheter is fixed

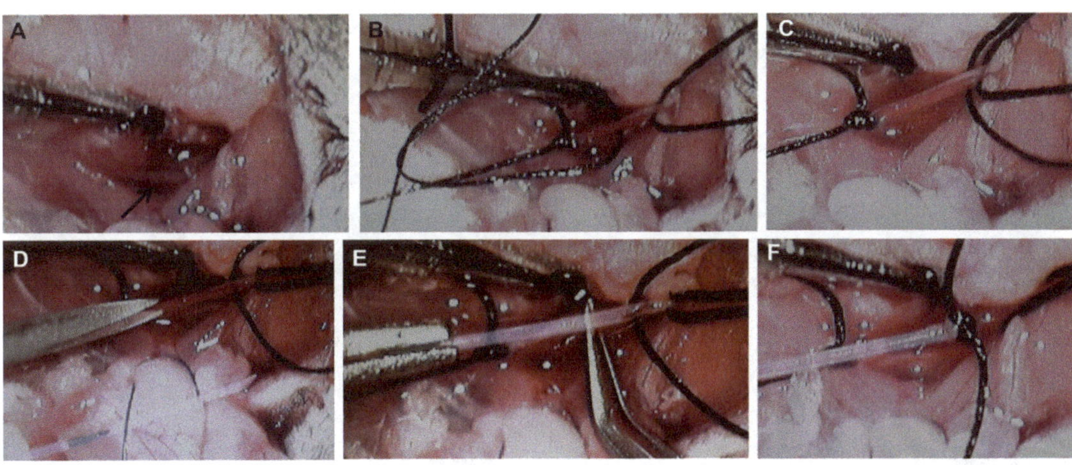

Fig. 4 Placement of catheter in the right carotid artery. (**a**) Exposure of carotid (arrow). (**b**) Suture placement. (**c**) Sutures tied off. (**d**) Incision. (**e**) Introduction of catheter. (**f**) Catheter secured in place. Note: A left-ventricular catheter with pressure and conductance sensors could be introduced instead

5. Tie the distal ligation (further from the animal's heart) using a surgeon's knot (*see* Fig. 4b, c).
6. Hold the proximal ligation (closer to the heart), to keep the artery shut (*see* **Note 17**).
7. Performing a slanted incision, open up the vessel between the ligations with microscissors (*see* Fig. 4d).

8. Ease the catheter into the artery and through the ligation (*see* **Notes 18, 19** and Fig. 4e).
9. Make sure to keep holding the catheter and ligation, then tie it off with pressure (*see* **Note 20** and Fig. 4f).
10. Flush the catheter to ensure patency.
11. Hook up the catheter to a hemodynamic monitor.

3.5 Bladder Catheter (See Note 21 and Fig. 5)

1. Perform a midline laparotomy (0.7 mm) above the bladder.
2. Puncture the bladder and insert the catheter.

3.6 Portal Vein Flow (1.5SL)

The following procedures are optional (*see* **Note 22** and Fig. 6).

1. Perform a transverse laparotomy.
2. Identify and expose the portal vein (*see* **Note 23**).
3. Place the perivascular flow probe around the vessel (*see* Fig. 6 and **Note 24**).
4. Connect the probe to a multiple-channel ultrasonic transit-time flowmeter.

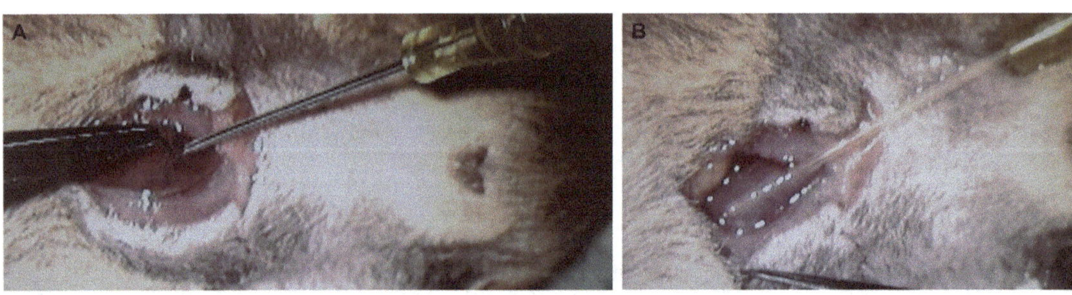

Fig. 5 Bladder catheter placement. (**a**) Exposure and puncture of bladder. (**b**) Bladder catheter in place

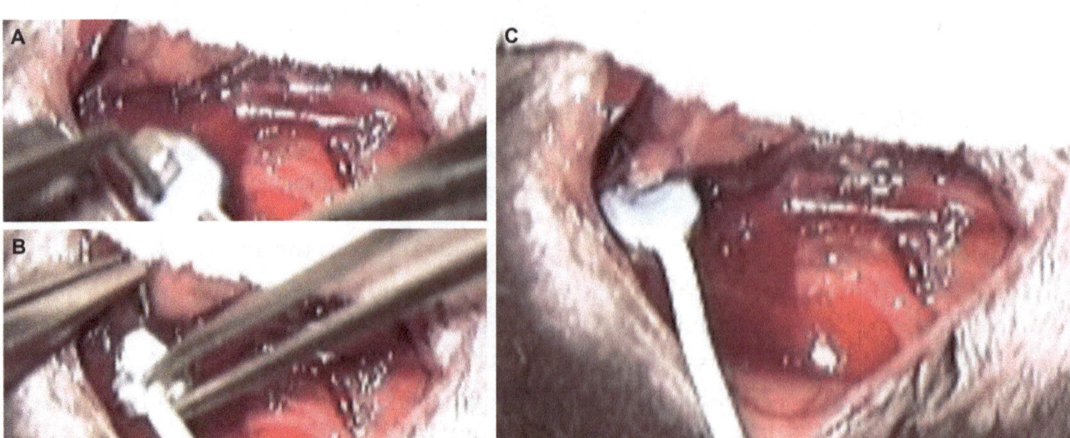

Fig. 6 Portal vein flow probe. (**a**) Exposure of the portal vein. (**b**) Placement of the flow probe. (**c**) Flow probe in place

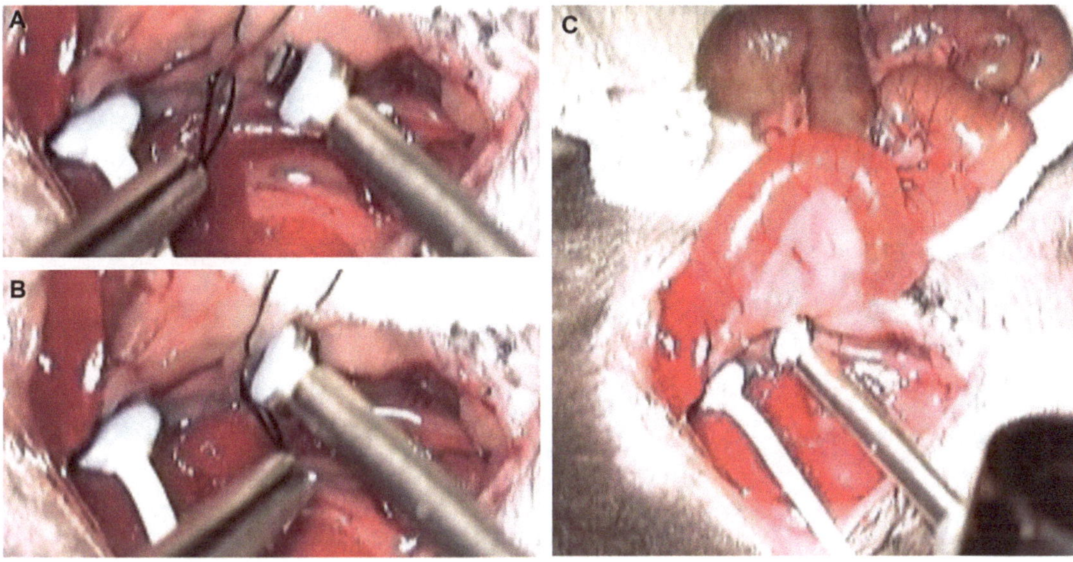

Fig. 7 Superior mesenteric artery (SMA) flow probe. (**a**) Exposure of the SMA. Placing of sutures. (**b**) Placement of the flow probe. (**c**) Flow probe in place and sutures removed. Place back intestines

3.7 Superior Mesenteric Artery (SMA) Flow Probe (0.7PSB)

This procedure is optional (*see* Fig. 7).

1. Identify the SMA (*see* **Note 22**).
2. Prepare two ligations around the SMA (*see* jugular vein cannulation, Fig. 7a).
3. Place the perivascular flow probe and connect to a multiple-channel ultrasonic transit-time flowmeter.
4. Remove sutures used for ligations.
5. Carefully place back the intestines after placing probes (*see* **Note 25**). More specific information about our flow measurements can be found in a prior report [13].

3.8 Patient Management

1. The basic MICU setup can be seen in Fig. 8.
2. To prevent distress to the animal, it is imperative to maintain general anesthesia throughout the experiment to guarantee complete tolerance against noxious stimuli. This should be accomplished by continuous intravenous administration of ketamine, midazolam, and fentanyl to maintain deep sedation.
3. Hemodynamics, lung mechanics, blood gas tensions, acid–base status, glycemia, and lactatemia and body temperature should be recorded hourly.
4. Blood draws for blood gas analysis are easily performed by removing 30 μl of blood from the arterial catheter with a glass capillary.

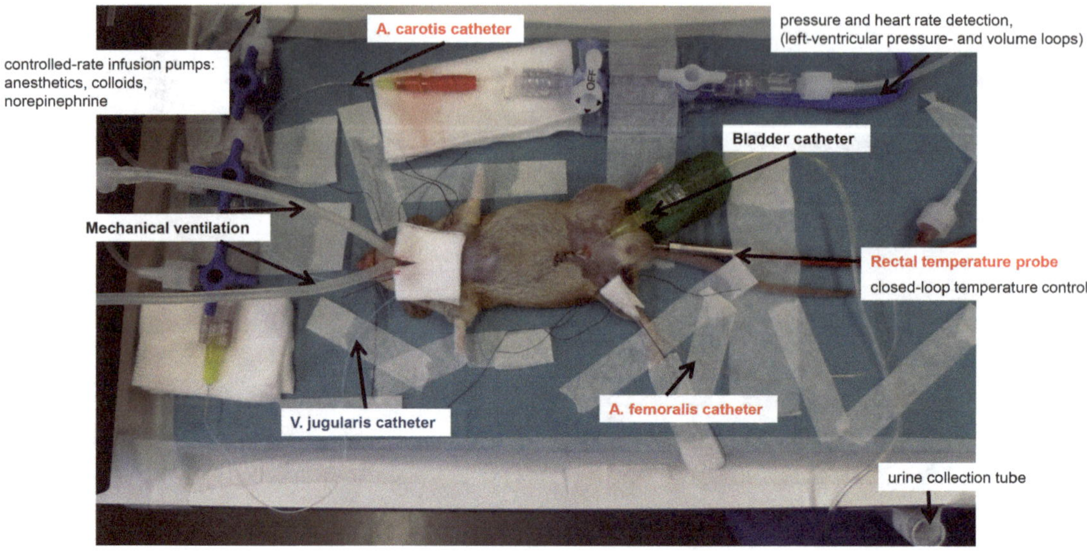

Fig. 8 Mouse in the ICU. The incision over the bladder is stitched up. The surgical site on the throat is kept open but covered with gauze. Note: A femoral artery catheter can be used for blood sampling, if a left-ventricular catheter is used in the carotid

5. Cold stress can occur in mice under deep anesthesia and/or as a natural response to injury [28]. To prevent this, the maintenance of the body temperature with a controlled closed-loop system is recommended.
6. Normotensive hemodynamics (i.e., mean arterial pressure (MAP) $\geq$ 55 mmHg) should be maintained by continuous i.v. colloid fluids (10 μl g^{-1} h^{-1}) and norepinephrine infusion in a balanced electrolyte solution, based on need. The appropriate target blood pressure should be chosen based on the anesthesia regimen and mouse strain, that is, for the protocol suggested here using C57BL6 mice, the literature suggests a typical MAP $\geq$55 mmHg [12].
7. The initial ventilator settings suggested in this protocol are: $F_iO_2 = 0.21$, respiratory rate 150 per min, tidal volume 4–6 μl/g, inspiratory/expiratory time ratio 1:2, and positive end-expiratory pressure (PEEP) 3 cmH_2O [24, 25, 29]. Following ARDS guidelines, mechanical ventilation during the experiment should be adjusted according to blood gas tensions: the respiratory rate should be modified to maintain an arterial partial pressure of CO_2 (P_aCO_2) between 30 and 40 mmHg, the PEEP should be varied according to the arterial P_aO_2 (if $P_aO_2/F_iO_2 > 300$ mmHg: PEEP = 3 cmH_2O; $P_aO_2/F_iO_2 < 300$ mmHg: PEEP = 5 cmH_2O; $P_aO_2/F_iO_2 < 200$ mmHg: PEEP = 8 cmH_2O) [24, 25, 29].

8. Recurrent recruitment maneuvers (5 s hold at 18 cmH_2O, at least hourly) should be performed to prevent anesthesia- and/or supine position-induced atelectasis with subsequent reduction of thoracopulmonary compliance. Additionally, recording plasma and urine creatinine concentrations together with the measurement of urinary output allows for calculation of creatinine clearance to obtain information on kidney function.

4 Notes

1. Knowing the weight of the animal is crucial to determine the dose of anesthetics to be used at the start of the experiment, and to determine ventilator settings later on.
2. Add approximately 200–300 μl Sevoflurane to a paper towel on the bottom of a chamber, put the animal into the chamber and wait for it to fall asleep. It is important to have a chamber made out of see-through material (such as glass) to observe the effects of the anesthesia on the animal.
3. Be careful to inject the anesthetic solution into the peritoneal cavity without injecting an abdominal organ. The maximum volume administered should not exceed 10 μl/g body weight.
4. Hair clipping will make the surgical procedures easier to perform.
5. Eye ointment is important to keep the eyes from drying out, and may need to be reapplied during the experiment.
6. Make sure to lubricate the thermometer before insertion, to prevent injury to the animal. Managing the animal's temperature is crucial to translational medicine. Rodents, in particular, are prone to cold stress under anesthesia.
7. Performing a tracheostomy is the most convenient way to provide lung-protective mechanical ventilation. Mechanical ventilation allows for deeper anesthesia and sedation with a very low risk for respiratory insufficiency, alleviating stress to the animal. Repetitive recruitment maneuvers prevent atelectasis. The management of the ventilator is described in Subheading 3.8.
8. Blunt dissection of muscle tissue by pulling fibers apart instead of cutting reduces bleeding (muscle tissue is highly vascularized).
9. The incision on the trachea can either be performed with microsurgical scissors or a 20-gauge cannula.
10. The initial recruitment maneuver will allow you to assess baseline lung function and ensure proper subsequent mechanical

ventilation. This initial maneuver should be performed in the same way as the following recurrent recruitment maneuvers (described in Subheading 3.8, **step 8**).

11. Be careful to not injure the vessel—its wall is very thin. Dissect tissue bluntly. Do not grab the vein with forceps. Prevent drying of the tissue by dripping isotonic saline on the tissue with a syringe.
12. The incision must not be bigger than the diameter of the cannula you want to use.
13. Prefill catheters, hooked up to a 1 ml syringe, with isotonic heparinized NaCl solution. Prepare catheters of various sizes to be ready to use the most appropriate size for the animal.
14. Intravenous administration of anesthesia is easy to control and adjust. For a setup without mechanical ventilation, anesthesia should instead be performed by repetitive i.p. injections, as described earlier, and sevoflurane inhalation, since the anesthesia depth by i.v. administration could lead to respiratory insufficiency and is not manageable without mechanical ventilation. We do not advise this approach, since the risk for respiratory insufficiency is still high. Mechanical ventilation allows for deeper anesthesia and potentially alleviates stress to the animal.
15. Prepare as you did for the cannulation of the jugular vein. Instead of a regular catheter, a left-ventricular catheter can optionally be inserted into the carotid artery to allow for the detection of left-ventricular pressure–volume loops to determine left-ventricular function.
16. The carotid artery lies underneath the muscular layer, close to the thin internal jugular vein and the vagus nerve. Be careful not to injure any vessel or nerve.
17. There is pressure on the artery, so make sure to keep it shut before performing the incision.
18. Blood should now shoot into the cannula.
19. A 1.4-Fr left-ventricular catheter (with pressure and conductance sensors) has to be inserted deeper than a regular catheter: advance into the heart across the aortic valve under the guidance of the online pressure signal. Further information on our specific left-ventricular catheterization and related measurements can be found in Albuszies et al. [13].
20. Do not let go of the catheter before it is tied into place.
21. Catheterization of the bladder allows for urine collection and analysis. Direct catheterization alleviates difficulties of transurethral catheterization due to sex-specific anatomical differences. Instant adhesive can be used to immobilize the catheter on the animal's flank.

22. Vascular flow probes can be used to determine hepatic microvascular perfusion, capillary hemoglobin concentration, and oxygen saturation.
23. Use a moist cotton swab to move the intestines out of the way. They should be placed onto the skin of the animal next to the incision onto a piece of sterile gauze (soaked in sterile saline) and covered up to prevent drying.
24. The ligations can be used to lift up the vessel to facilitate the placing of the flow probe.
25. Use a moist cotton swab.

Acknowledgments

Funded by the Deutsche Forschungsgemeinschaft (DFG), grant number Project-ID 251293561—Collaborative Research Center (CRC) 1149 to PR and TM, and the DFG-funded Graduiertenkolleg "PulmoSens" GRK 2203 to PR.

References

1. Traber DL (1999) Expired nitric oxide and shock in higher order species. Crit Care Med 27(2):255–256
2. Osuchowski MF, Ayala A, Bahrami S, Bauer M, Boros M, Cavaillon JM, Chaudry IH, Coopersmith CM, Deutschman CS, Drechsler S, Efron P, Frostell C, Fritsch G, Gozdzik W, Hellman J, Huber-Lang M, Inoue S, Knapp S, Kozlov AV, Libert C, Marshall JC, Moldawer LL, Radermacher P, Redl H, Remick DG, Singer M, Thiemermann C, Wang P, Wiersinga WJ, Xiao X, Zingarelli B (2018) Minimum quality threshold in pre-clinical sepsis studies (MQTiPSS): an international expert consensus initiative for improvement of animal modeling in sepsis. Shock 50(4):377–380
3. Hellman J, Bahrami S, Boros M, Chaudry IH, Fritsch G, Gozdzik W, Inoue S, Radermacher P, Singer M, Osuchowski MF, Huber-Lang M (2019) Part III: minimum quality threshold in preclinical sepsis studies (MQTiPSS) for fluid resuscitation and antimicrobial therapy endpoints. Shock 51(1):33–43
4. Angus DC, van der Poll T (2013) Severe sepsis and septic shock. N Engl J Med 369 (9):840–851
5. Hollenberg SM, Dumasius A, Easington C, Colilla SA, Neumann A, Parrillo JE (2001) Characterization of a hyperdynamic murine model of resuscitated sepsis using echocardiography. Am J Respir Crit Care Med 164(5):891–895
6. Guillon A, Preau S, Aboab J, Azabou E, Jung B, Silva S, Textoris J, Uhel F, Vodovar D, Zafrani L, de Prost N, Radermacher P (2019) Preclinical septic shock research: why we need an animal ICU. Ann Intensive Care 9(1):66
7. Zhou F, Peng ZY, Bishop JV, Cove ME, Singbartl K, Kellum JA (2014) Effects of fluid resuscitation with 0.9% saline versus a balanced electrolyte solution on acute kidney injury in a rat model of sepsis. Crit Care Med 42(4): e270–e278
8. Semler MW, Kellum JA (2019) Balanced crystalloid solutions. Am J Respir Crit Care Med 199(8):952–960
9. Marshall JC, Bosco L, Adhikari NK, Connolly B, Diaz JV, Dorman T, Fowler RA, Meyfroidt G, Nakagawa S, Pelosi P, Vincent JL, Vollman K, Zimmerman J (2017) What is an intensive care unit? A report of the task force of the world Federation of Societies of intensive and critical care medicine. J Crit Care 37:270–276
10. Dellinger RP, Carlet JM, Masur H, Gerlach H, Calandra T, Cohen J, Gea-Banacloche J, Keh D, Marshall JC, Parker MM, Ramsay G, Zimmerman JL, Vincent JL, Levy MM (2004) Surviving sepsis campaign guidelines for

management of severe sepsis and septic shock. Crit Care Med 32(3):858–873

11. Singer M, Deutschman CS, Seymour CW, Shankar-Hari M, Annane D, Bauer M, Bellomo R, Bernard GR, Chiche JD, Coopersmith CM, Hotchkiss RS, Levy MM, Marshall JC, Martin GS, Opal SM, Rubenfeld GD, van der Poll T, Vincent JL, Angus DC (2016) The third international consensus definitions for sepsis and septic shock (Sepsis-3). JAMA 315 (8):801–810
12. Zuurbier CJ, Emons VM, Ince C (2002) Hemodynamics of anesthetized ventilated mouse models: aspects of anesthetics, fluid support, and strain. Am J Physiol Heart Circ Physiol 282(6):H2099–H2105
13. Albuszies G, Radermacher P, Vogt J, Wachter U, Weber S, Schoaff M, Georgieff M, Barth E (2005) Effect of increased cardiac output on hepatic and intestinal microcirculatory blood flow, oxygenation, and metabolism in hyperdynamic murine septic shock. Crit Care Med 33(10):2332–2338
14. Brealey D, Brand M, Hargreaves I, Heales S, Land J, Smolenski R, Davies NA, Cooper CE, Singer M (2002) Association between mitochondrial dysfunction and severity and outcome of septic shock. Lancet 360 (9328):219–223
15. Vincent JL, De Backer D (2013) Circulatory shock. N Engl J Med 369(18):1726–1734
16. Merz T, Vogt JA, Wachter U, Calzia E, Szabo C, Wang R, Radermacher P, McCook O (2017) Impact of hyperglycemia on cystathionine-γ-lyase expression during resuscitated murine septic shock. Intensive Care Med Exp 5(1):30
17. Francis RC, Vaporidi K, Bloch KD, Ichinose F, Zapol WM (2011) Protective and detrimental effects of sodium sulfide and hydrogen sulfide in murine ventilator-induced lung injury. Anesthesiology 115(5):1012–1021
18. Dreyfuss D, Saumon G (1998) Ventilator-induced lung injury: lessons from experimental studies. Am J Respir Crit Care Med 157 (1):294–323
19. de Prost N, Ricard JD, Saumon G, Dreyfuss D (2011) Ventilator-induced lung injury: historical perspectives and clinical implications. Ann Intensive Care 1(1):28
20. Dreyfuss D, Soler P, Basset G, Saumon G (1988) High inflation pressure pulmonary edema. Respective effects of high airway pressure, high tidal volume, and positive end-expiratory pressure. Am Rev Respir Dis 137(5):1159–1164
21. Turnbull IR, Wlzorek JJ, Osborne D, Hotchkiss RS, Coopersmith CM, Buchman TG (2003) Effects of age on mortality and antibiotic efficacy in cecal ligation and puncture. Shock 19(4):310–313
22. Hartmann C, Hafner S, Scheuerle A, Möller P, Huber-Lang M, Jung B, Nubaum B, McCook O, Gröger M, Wagner F, Weber S, Stahl B, Calzia E, Georgieff M, Szabó C, Wang R, Radermacher P, Wagner K (2017) The role of cystathionine-γ-Lyase in blunt chest trauma in cigarette smoke exposed mice. Shock 47(4):491–499
23. Wagner K, Gröger M, McCook O, Scheuerle A, Asfar P, Stahl B, Huber-Lang M, Ignatius A, Jung B, Duechs M, Möller P, Georgieff M, Calzia E, Radermacher P, Wagner F (2015) Blunt chest trauma in mice after cigarette smoke-exposure: effects of mechanical ventilation with 100% O_2. PLoS One 10(7): e0132810
24. Wepler M, Merz T, Wachter U, Vogt J, Calzia E, Scheuerle A, Möller P, Gröger M, Kress S, Fink M, Lukaschewski B, Rumm G, Stahl B, Georgieff M, Huber-Lang M, Torregrossa R, Whiteman M, McCook O, Radermacher P, Hartmann C (2019) The mitochondria-targeted H2S-donor AP39 in a murine model of combined hemorrhagic shock and blunt chest trauma. Shock 52(2):230–239
25. Hartmann C, Gröger M, Noirhomme JP, Scheuerle A, Möller P, Wachter U, Huber-Lang M, Nussbaum B, Jung B, Merz T, McCook O, Kress S, Stahl B, Calzia E, Georgieff M, Radermacher P, Wepler M (2019) In-depth characterization of the effects of cigarette smoke exposure on the acute trauma response and hemorrhage in mice. Shock 51(1):68–77
26. McCook O, Georgieff M, Scheuerle A, Möller P, Thiemermann C, Radermacher P (2012) Erythropoietin in the critically ill: do we ask the right questions? Crit Care 16 (5):319
27. Angus DC, Linde-Zwirble WT, Lidicker J, Clermont G, Carcillo J, Pinsky MR (2001) Epidemiology of severe sepsis in the United States: analysis of incidence, outcome, and associated costs of care. Crit Care Med 29 (7):1303–1310
28. Iskander KN, Osuchowski MF, Stearns-Kurosawa DJ, Kurosawa S, Stepien D, Valentine C, Remick DG (2013) Sepsis: multiple abnormalities, heterogeneous responses, and evolving understanding. Physiol Rev 93 (3):1247–1288

29. Gröger M, Wepler M, Wachter U, Merz T, McCook O, Kress S, Lukaschewski B, Hafner S, Huber-Lang M, Calzia E, Georgieff M, Nagahara N, Szabó C, Radermacher P, Hartmann C (2019) The effects of genetic 3-Mercaptopyruvate Sulfurtransferase deficiency in murine traumatic-hemorrhagic shock. Shock 51(4):472–478

Chapter 12

Creation of BLT Humanized Mice for Sepsis Studies

Erica L. Heipertz and Wendy E. Walker

Abstract

Mice are a suitable animal model for sepsis studies because they recapitulate many aspects of the pathophysiology observed in septic human patients. It is ethically preferable to use mice for research over higher sentient species, when scientifically appropriate. Mice are also advantageous for research due to their small size, modest housing needs, the availability of genetically modified strains, and the broad range of reagents available for scientific assays on this species. Nevertheless, there are some intrinsic differences between mice and humans that should be recognized when considering the translational potential of sepsis therapies. It is often wise to complement traditional mouse studies with animal models that exhibit even greater similarity to humans, and in particular, models that better recapitulate the human immune response. Humanized mice are a promising tool to bridge this interspecies research gap. Herein, we provide a protocol to generate BLT humanized mice and describe their sepsis phenotype after cecal ligation and puncture (CLP).

Key words Humanized mice, BLT, Sepsis

1 Introduction

Sepsis continues to be the leading cause of in-hospital deaths. Despite carefully coordinated treatments, the mortality rate in sepsis remains high [1, 2]. Novel therapies that modulate the immune response may improve treatment strategies and thus mortality of sepsis. Because sepsis arises from complex pathophysiological processes involving multiple organ systems, in vitro assays are not adequate to fully recapitulate this condition. Therefore, in vivo models are required to study sepsis.

Mice are the most commonly used species to study sepsis. However, recent reports have generated debate over whether mouse models adequately recapitulate human sepsis [3–6]. One transcriptome analysis suggested that there are major differences in the gene expression patterns induced in septic mice vs. humans [4]. In contrast, a subsequent report used different analysis parameters and found a better correlation between gene expression patterns in the two species [3]. Regardless of gene expression patterns, septic mice and humans share numerous physiological

Wendy E. Walker (ed.), *Sepsis: Methods and Protocols*, Methods in Molecular Biology, vol. 2321,
https://doi.org/10.1007/978-1-0716-1488-4_12,

and pathophysiological traits [5], and mouse research still has a strong potential to inform our understanding of sepsis. Nevertheless, there are some intrinsic differences between the immune systems of mice and humans that may influence disease pathogenesis, and these deserve careful consideration when developing new therapies for sepsis.

Humanized mice are a valuable tool that can help bridge the interspecies research gap. Genetically altered mice have been generated that lack most immune cell subsets, and express cytokines to support human immune cell development. In the BLT model, NOD/SCID/IL2Rγ−/− (NSG) strain mice are irradiated and small fragments of fetal liver and thymus are injected under their kidney capsule. Subsequently, the mice are intravenously injected with CD34+ cells isolated from the liver. This procedure results in reconstitution of the leukocyte compartment with human cells within 12–15 weeks. The transplanted stem cells give rise to nearly all lineages of the human immune system, including T cells, B cells, NK cells, monocytes, macrophages, and dendritic cells (DCs) [7–9]. A thymic organoid develops in the kidney, providing a site for the T cells to be educated against human HLA, this confers the ability for T cells to recognize antigen presented by human cells, and leads to improved reconstitution of T cells [7–9].

CLP induces sepsis in humanized BLT mice, in a similar manner to inbred mouse strains (100% mortality by 3 days, using a severe CLP protocol) [10]. In contrast, unreconstituted NSG strain mice are resistant to CLP-sepsis [10]. Unreconstituted mice possess mouse neutrophils that control the bacterial infection, however they do not possess macrophages or DCs, and do not produce abundant cytokines in response to the bacterial infection [10]. These results are consistent with the notion that a human cytokine storm induces sepsis in BLT mice subject to CLP. Additional reports have used humanized mice to study sepsis with similar success, employing the BLT model [11, 12] and closely related models of NSG mice transplanted with cord blood CD34 + cells [11–15]. In this chapter we describe the method to create BLT humanized mice. This procedure can also be adapted to reconstitute NSG mice with cord blood CD34+ cells, by skipping the kidney capsule implantation and by using mononuclear cells prepared from cord blood instead of fetal liver for CD34+ cell enrichment.

2 Materials

2.1 Experimental Animals and Biological Samples

1. NOD/SCID/IL2Rgamma−/− (NSG) mice, male and/or female, aged 6–10 weeks (*see* **Note 1**).
2. Second trimester (17–20 week) fetal liver and thymus specimens (*see* **Note 2**).

2.2 Isolation of Mononuclear Cells from the Fetal Liver

1. Biosafety cabinet.
2. Ice bucket.
3. Sterile petri dishes.
4. Dulbecco's phosphate buffered saline without calcium chloride or magnesium chloride (PBS).
5. RPMI containing 1% penicillin–streptomycin.
6. Digestion buffer: RPMI containing 1% penicillin–streptomycin, 1 U/mL DNase I, and 2 mg/mL collagenase–dispase.
7. Forceps.
8. Scissors.
9. 15 and 50 mL polypropylene tubes.
10. 40 μm cell strainers.
11. 1 mL pipetman.
12. 1 mL pipette tips with the end cut off to widen the diameter, sterile.
13. Water bath, set to 37 °C.
14. Ficoll-Paque, sterile, endotoxin-tested.
15. Refrigerated centrifuge capable of 400 × *g* force, with rotors to hold 15 and 50 mL tubes.
16. Hemocytometer slide.

2.3 Magnetic Enrichment of CD34+ Cells

1. CD34 MicroBead Kit UltraPure, human (Miltenyi Biotec).
2. MACS buffer: PBS (pH 7.2) 0.5% bovine serum albumin, and 2 mM EDTA.
3. MidiMACS or QuadroMACS separator magnet, with MACS MultiStand (Miltenyi Biotec) (*see* **Note 3**).
4. LS MACS columns (Miltenyi Biotec).
5. Optionally: Human Fc blocking solution, fluorescently conjugated anti-human CD34 antibody, FACS staining buffer (PBS, 2% fetal bovine serum), fixation buffer (2% paraformaldehyde in PBS), round bottom 5 mL polypropylene tubes and a flow cytometer to determine CD34+ cell purity (*see* **Note 4**).

2.4 Irradiation

1. X-ray irradiator.
2. Allentown cages or other container suitable for the irradiator.

2.5 Kidney Capsule Implantation

1. Sterile petri dishes.
2. Isoflurane vaporizer.
3. Isoflurane.
4. Alcohol prep pads.

5. Buprenorphine SR.
6. 0.5 cm^3 insulin syringes, 28G × 1/2 in. needle.
7. Sharps bin.
8. Clippers, clipper disinfectant and lubricant spray.
9. Eye lubricant.
10. Betadine surgical scrub (povidone–iodine, 7.5%).
11. 70% ethanol in dH_2O (optional, alcohol prep pads can be used instead).
12. Cotton swabs.
13. Two heating pads for thermal support (*see* **Note 5**).
14. Surgical drape: large pieces to cover the benchtop (optional) and small 25 cm squares (sterile).
15. Press-and-seal food wrap, cut into 15 cm pieces, with a 2.5 cm circular hole cut in the center of each piece (*see* **Note 6**).
16. Sterile surgical tools: surgical scissors, curved and straight tissue forceps (for handling skin during the incision and suturing process), extra delicate micro dissecting forceps; serrated, with a slight curve, 0.5 mm tip width × 4″ length (for lifting the kidney capsule); Dumont tweezers style 4, 0.02 × 0.06 mm (for puncturing the kidney capsule), Delicate spring scissors (for chopping up the liver and thymus tissue into fragments) (*see* **Note 7**).
17. PBS (as described in Subheading 2.2).
18. 0.5 mL syringe.
19. Blunt needles, 0.5 in. length, 20G.
20. Absorbable surgical suture with needle attached, sterile (such as coated Vicryl, 27 in., 5-0 with RB-1 17 mm 1/2c tapered needle).
21. Normal saline or lactated Ringer's solution in a 50 mL tube.
22. Heat block with insert block for a 50 mL tube, set to 37 °C.
23. Imipenem–cilastatin antibiotic solution (4 mg/mL, suspended in normal saline or lactated Ringer's solution).
24. Benchtop bead sterilizer (*see* **Note 7**).

2.6 Tail Vein Injection

1. Isoflurane machine (optional, *see* **Note 8**).
2. Mouse restraint device (optimal, *see* **Note 8**).
3. Heat block with insert block for a 50 mL tube, set to 40 °C.
4. 50 mL tube filled with distilled water.
5. Paper towels.
6. Alcohol prep pads.

7. Platform, approximately 15–20 cm high (e.g., a plastic box).
8. 0.5 cc insulin syringes, 28G × 1/2 in. needle.
9. Sharps bin.
10. Gauze pads.

2.7 Staining Mouse PBMCs to Verify Leukocyte Reconstitution

1. Lancet or heparinized capillary tubes (microhematocrit), to collect blood.
2. Heparinized blood collection tubes.
3. 1.5 mL microcentrifuge tubes.
4. Histopaque-1077.
5. Ice bucket.
6. FACS staining buffer (PBS, 2% fetal bovine serum).
7. Human Fc blocking solution (optional, *see* **Note 9**).
8. Fluorescently labeled antibodies against human CD45, CD3, CD19, and CD14.
9. Optional: fixation buffer (2% paraformaldehyde in PBS).
10. Round bottom, 5 mL polypropylene tubes.
11. Flow cytometer.

3 Methods

3.1 Isolating Mononuclear Cells from the Fetal Liver

Please note: fetal tissue specimens are potentially infected with human pathogens, and may require Biosafety Level 2 handling precautions. Specimen testing for HIV and Hepatitis B can help to mitigate this risk. Contact your Institutional Biosafety Committee to determine appropriate handling procedures.

1. Prepare the sterile biosafety cabinet.
2. Obtain samples of fetal liver specimens from a second trimester (17-20 week) fetal cadaver (*see* **Note 2**).
3. Dissect 1/3 of the tissue and place it in a sterile petri dish containing PBS. Set this tissue aside along with the thymus tissue for kidney capsule implantation.
4. Place the remaining liver tissue in a petri dish with 2.5 mL of RPMI containing 1% penicillin–streptomycin.
5. Mince the tissue with the scissors.
6. Add 2.5 mL digestion buffer.
7. Transfer the tissue fragments to a 50 mL tube, using a 1 mL pipette with the tip cut off.
8. Incubate the sample for 1 h in a 37 °C water bath to dissociate the tissue. Agitate the sample every 5–10 min.

9. Add an additional 2 mL of RPMI containing 1% penicillin–streptomycin to the sample.
10. Filter the sample through a 40 μM cell strainer placed on top of a 50 mL tube. Discard the filter.
11. Prepare 5 mL Ficoll-Paque in a 15 mL tube.
12. Overlay the cell suspension on top of the Ficoll-Paque (*see* **Note 10**).
13. Spin the sample at 400 × *g*, 30 min at room temperature with no brake.
14. Remove the sample from the centrifuge. The sample should appear as two layers, with a cloudy band at the interface. Using a 1 mL pipette, collect this cloudy band (which contains the mononuclear cells) and place the cells in a fresh 50 mL tube.
15. Add 45 mL PBS to wash the cells.
16. Spin the sample at 400 × *g* 10 min at room temperature (with normal braking) to pellet the cells.
17. Carefully aspirate the supernatant and resuspend the cell pellet in 10 mL PBS.
18. Count the cells, using a hemocytometer slide (*see* **Note 11**).
19. Spin the sample at 400 × *g* 10 min at room temperature (with normal braking) to pellet the cells.
20. Carefully aspirate the supernatant and resuspend the cells in 300 μL MACS buffer per 10^8 cells.
21. Optionally, take a 5–10 μL aliquot of these unfractionated cells for antibody staining and flow cytometry (*see* **Note 4**).

3.2 Magnetic Enrichment of CD34+ Cells

1. Prepare a solution of mononuclear cells from the fetal liver as described above.
2. Obtain the CD34 MicroBead UltraPure kit (Miltenyi Biotec), containing FcR blocking reagent and CD34 microbeads (*see* **Note 12**). Store the reagents on ice and mix by pipetting, prior to use.
3. Prepare the MACS buffer and store on ice during the enrichment process.
4. Add 100 μL FcR Blocking reagent per 10^8 cells.
5. Add 100 μL CD34 MicroBeads per 10^8 cells.
6. Mix the sample well and incubate at 4 °C (refrigerator) for 30 min. During this incubation, cool the centrifuge to 4 °C.
7. Add 5–10 mL MACS buffer per 10^8 cells to wash the sample. Centrifuge the tube at 300 × *g* for 10 min, 4 °C.
8. Carefully aspirate the supernatant and resuspend the pellet in 500 μL MACS buffer per 10^8 cells.

9. Prepare one LS column to separate the sample (*see* **Note 11**). Insert the column into the MidiMACS or QuadroMACS separator magnet attached to the MACS MultiStand. Ensure the column is fully pushed back into the holder. Place a 50 mL collection tube below the LS column. Work quickly through the following steps and do not allow the column to sit dry between steps.
10. Add 3 mL MACS buffer to prepare the column. Allow the solution to drip through.
11. Apply the cell suspension onto the column and allow to drip through.
12. Wash the column three times with 3 mL MACS buffer. Allow to drip through.
13. Once the final wash is complete, remove the column from the magnet and place it atop a 50 mL collection tube. Apply 5 mL MACS buffer to the column and insert the plunger and firmly push it down to flush the enriched CD34+ cells out of the column.
14. Count the enriched CD34+ cells using a hemocytometer slide (*see* **Note 11**).
15. Spin the sample at 300 × *g* 10 min at 4 °C.
16. Aspirate the supernatant and resuspend the cells in PBS at a concentration of 2.5×10^6/mL.
17. Optionally, remove a 5–10 μL aliquot of these enriched CD34+ cells for antibody staining and flow cytometry (*see* **Note 4**).

3.3 Mouse Irradiation

Before commencing any animal experiments, you must obtain approval from your Institutional Animal Care and Use Committee (IACUC), or for investigators outside of the USA, from the relevant body governing the ethical use of animals in your region.

1. Warm up the irradiator, according to the manufacturer's instructions (*see* **Note 13**).
2. Place the mice in a clean Allentown cage, or a restraint device suitable for your irradiator (*see* **Note 14**).
3. Place the cage or restraint device in the irradiator.
4. Program the irradiator to deliver a sublethal dose of 1–2 Gy.
5. Remove the mice from the irradiator and place them in a clean cage (*see* **Note 15**).

3.4 Kidney Capsule Implantation

1. Following the irradiation (on the same day), prepare mice for kidney capsule implantation.
2. Wash your hands. Don a surgical gown, mask, hair net and gloves.

3. Prepare the area: clean the surgical bench with disinfectant and cover the surface with surgical drape (optional). Designate separate areas of the workbench for preoperative, intraoperative, and postoperative use, and place the appropriate materials in each area. Prewarm the thermal support pads to 38 °C (one in the intraoperative area, one in the postoperative area). Preheat the bead sterilizer (to sterilize instruments between multiple animals) and the heat block containing the tube of normal saline or lactated Ringer's solution (37 °C). Prepare 15 cm squares of press-and-seal wrap and cut a 2.5 cm circle in the center of each piece. Prepare a recovery chamber (empty cage without litter) in the postoperative area, position it half on and half off the thermal support pad, so that the animals may move if they get too warm after regaining consciousness.
4. Place a mouse in the chamber of the precision isoflurane vaporizer to achieve anesthesia (*see* **Note 16**). Set the machine to an oxygen flow of 1 L/min, 3.0% isoflurane for induction and then adjust the isoflurane to 1.0–2.5% as necessary to maintain appropriate Stage III, plane 2 anesthesia (*see* **Note 17**). Adequate anesthetic depth should be confirmed prior to starting surgery (by monitoring heart rate, respiratory rate and lack of response to toe pinch) and monitored at 5-min intervals thereafter.
5. Weigh the animal and place it in the preoperative area, with its nose inserted into the nose cone of the isoflurane vaporizer. Inject the mouse with Buprenorphine SR (0.5–1 mg per kg body weight s.c.; *see* **Note 18**).
6. Shave the animal's left side, including a 2.5 cm square overlying the left kidney (*see* **Note 19** and Fig. 1a).
7. Apply veterinary eye lubricant to the animal's eyes to protect them from drying during surgery.
8. Scrub the surgical site with Betadine scrub. The solution should be applied to the center of the surgical area using sterile cotton-tipped swabs, then move outward in increasing circles from the center to the periphery. Next, repeat the scrub with an ethanol pad, or a swab dipped in 70% ethanol. Repeat the entire process three times, alternating between Betadine and ethanol. Use a fresh swab for each scrub.
9. Prepare a sterile square of surgical drape in the surgical area, on top of the thermal support pad.
10. Place the animal atop this drape, and position the nose cone of the isoflurane vaporizer on its nose, to maintain anesthesia. Position the mouse on its right side, so that the surgical site is on top.
11. Wash your hands. Put on sterile surgical gloves.

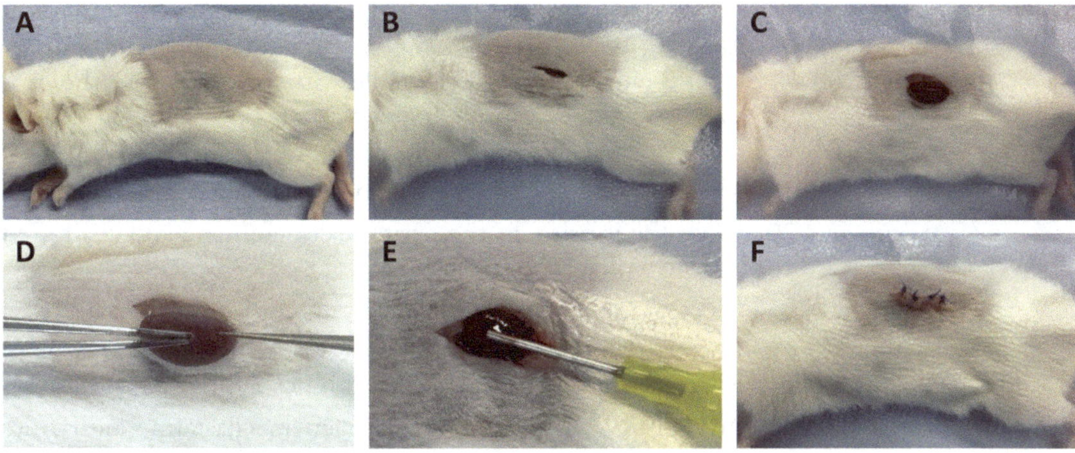

Fig. 1 Kidney capsule implantation surgery. (**a**) The surgical site is shaved. (**b**) The mouse is covered in press-and-seal translucent drape, with a hole cut in the center. A 1 cm incision is made in the skin and peritoneal membrane, overlying the kidney. (**c**) The kidney is extruded through the incision and lodged in place. (**d**) The kidney capsule is gently lifted and a hole is created with sharp forceps. (**e**) A blunt needle is inserted under the kidney capsule and fragments of tissue are injected. (**f**) The wound is closed in two layers with suture

12. Cover the animal with the press-and-seal wrap, positioning the hole over the surgical site (the translucent drape can be faintly visualized in Fig. 1b). Ensure you can visualize the animals' breathing and that it is unimpeded by the drape. Reposition the press-and-seal wrap around the left lower foot, to facilitate a toe pinch. Perform a toe pinch to ensure the animal is unresponsive to pain prior to starting surgery (*see* **Note 17**).
13. Using surgical scissors, cut a 1 cm long vertical incision in the skin of the abdomen overlying the left kidney (*see* Fig. 1b).
14. Subsequently, cut a similar incision in the underlying peritoneal membrane and associated musculature.
15. Prepare the kidney and liver fragments. Mix small pieces of fetal liver and thymus in a 1:1 ratio in a petri dish and moisten them with PBS. Mince the tissues into tiny fragments with micro spring scissors. Apply the blunt needle on the 1 mL syringe and draw 0.5 mm^3 of this mixed tissue into the needle into the needle. Set aside.
16. Locate the left kidney. Manipulate the peritoneal membrane incision around the kidney, to gently extrude it through the incision (*see* Fig. 1c). If the incision size is made appropriately, the kidney can be lodged within the hole to secure the organ in an accessible position.
17. After exposing the front of the kidney completely, gently lift the renal capsule with the fine curved forceps and puncture it with the sharp, straight forceps to create a tiny hole (*see* Fig. 1d). Do not let go of the kidney capsule as it will be difficult to relocate the hole.

18. Insert the blunt needle through this hole, under the renal capsule (*see* Fig. 1e) and release the tissue from the syringe. Gently remove the blunt needle without destroying the capsule. The kidney capsule incision is left open and will heal by itself.
19. Manipulate the kidney back into its normal position.
20. Close the incision in the peritoneal membrane and associated muscle with a running suture.
21. Close the skin incision with interrupted sutures (*see* Fig. 1f).
22. Inject the animal with prewarmed normal saline or lactated Ringer's solution s.c. (37 °C; 5 mL per 100 g body weight).
23. Administer one shot of imipenem–cilastatin antibiotics i.p. (25 mg/kg body weight) to prevent infection during the early postoperative period (*see* **Notes 20** and **21**).
24. Place the mouse in the recovery chamber on top of the thermal support pad, in the postoperative area.
25. Monitor the mouse at 5-min intervals until it is conscious and walking normally. Check the incision is intact with no bleeding. If bleeding occurs, apply pressure until it has stopped. Return the animals to a clean cage. Place moist chow on the cage floor to provide support postoperatively.

3.5 Injection of CD34+ Cells

1. Animals should be injected with CD34+ cells that are freshly isolated from the same cadaveric donor as the kidney and liver fragments. This injection may be performed on the same day as kidney capsule implantation, or on the following day (within 24 h of irradiation).
2. Wash your hands. Don a gown, mask, and gloves.
3. Prepare the work area: cover the box with paper towels and prepare the heat block, ice bucket containing the CD34+ cells, ethanol pads, and insulin syringes close by. Switch on the heat block and program it to 40 °C. Fill a 50 mL tube with water and place it in the heat block.
4. Tail vein injection can be performed with the animal sedated under isoflurane. If this method is to be used, place the animal in the isoflurane device. Set the machine to an oxygen flow of 1 L/min, 3.0% isoflurane for induction and then adjust the isoflurane to 1.0–2.5% as necessary to maintain appropriate Stage III, plane 2 anesthesia. Adequate anesthetic depth should be confirmed by monitoring heart and respiratory rate and lack of pedal reflex upon toe pinch.
5. Alternatively, the animal can be injected without sedation using a restraint device. In this case, load the mouse into the restraint device by grasping it firmly by the tail and manipulating it

backward, gently easing it into the tube-shaped restraint device. The tail should be manipulated into the opening at the back of the device. Insert the front attachment and adjust the position to restrain the mouse firmly without compressing the mouse (*see* **Note 22**). Ensure the mouse can breathe.

6. Load a 500 μL insulin syringe with 200 μL volume containing 1–5 × 10^5 CD34+ cells (*see* **Note 23**).
7. Open the 50 mL falcon of warm water and manipulate the mouse tail into the warm water. Approximately 30 s should be sufficient to dilate the blood vessels.
8. Place the mouse on its side atop the platform. If isoflurane is used, manipulate the nose cone into the correct position. If necessary, secure the nose cone in position with tape.
9. Locate the tail vein, running down the side of the mouse tail. Hold the tail taught supporting it with the fingers and thumb. Insert the needle, keeping it parallel to the vein as closely as possible. Inject the cells (*see* **Note 24**).
10. Following the intravenous injection, apply pressure to the injection site with a gauze pad for 2 min to stop the bleeding. Ensure all bleeding has stopped before returning the mouse to its cage. If isoflurane is used, allow the mouse to regain consciousness before returning the animal to its cage.

3.6 Animal Monitoring After Surgery and Injection

1. Monitor the animals daily for 14 days after surgery, to check they remain active and do not experience postoperative complications or signs of infection (*see* **Note 20**). Animals that experience continued signs of pain or discomfort should be given additional doses of analgesics (Buprenorphine SR, equal to the original dose, at 48–72 h intervals).Ensure that sutures remain intact and that the wound heals normally.
2. Check that no damage has occurred to the tail tissue as a consequence of the intravenous injection. If any issues arise contact the institutional veterinarian for advice.
3. Skin sutures should be removed at 7–10 days, once the wound has adequately healed.

3.7 Staining Mouse PBMCs to Verify Reconstitution

1. Bleed the animals at 12–15 weeks after tissue implantation and injection. Blood may be collected from the submandibular vein, using a lancet, or via retro-orbital bleed, using a heparinized capillary tube (a detailed protocol is provided in our prior methods chapter [16]).
2. Collect 60–80 μL blood into a heparinized blood collection tube.
3. Aliquot 500 μL Histopaque-1077 reagent into a set of 1.5 mL microcentrifuge tubes (one per mouse).

4. Layer the collected blood on top of the Histopaque-1077.
5. Spin the tubes at 400 × *g* for 2 min at room temperature in a benchtop centrifuge, with no braking. Collect PBMCs from the interface (*see* **Note 25**).
6. Add 500 μL FACS staining buffer and spin the sample at 650 × *g* for 4 min. Remove the supernatant and resuspend the cell pellet in the required volume of FACS staining buffer, typically 40–90 μL (*see* **Note 26**). Pipet up and down and gently vortex the sample to resuspend.
7. Optional: add human FC blocking reagent (diluted according to the manufacturer's instructions) and incubate the samples at room temperature for 10 min (*see* **Note 9**).
8. Create an antibody cocktail, by mixing antibodies against human CD45, CD19, CD3 and CD14 in FACS staining buffer (dilute antibodies according to the manufacturer's instructions). Add an appropriate volume to the tube of cells—typically 5–20 μL (*see* **Note 26**).
9. Incubate the samples on ice for 30 min.
10. Wash the cells with 500 μL staining buffer. Spin the samples at 650 × *g* for 2 min. Remove the supernatant.
11. Optional: the cells can be fixed by adding 200 μL fixation buffer and incubating on ice for 5–10 min. Subsequently, repeat the wash step described above.
12. Resuspend the cell pellet in 250 μL FACS buffer.
13. Analyze the cells with a flow cytometer. Detailed procedures for cell and singlet gating are described elsewhere (*see* reference [17] or Chap. 17 of this book). Typical results for reconstitution of BLT mice with human immune cells are shown in Fig. 2.

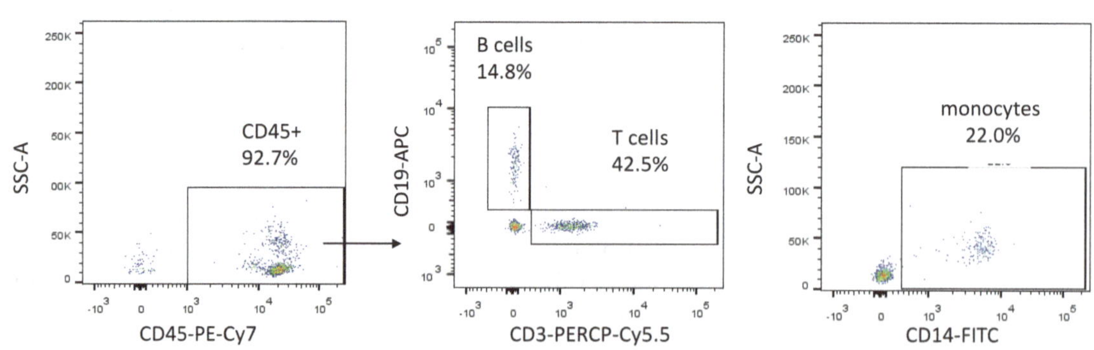

Fig. 2 Leukocyte populations in reconstituted BLT mice. We obtained peripheral blood from BLT mice 15 weeks after kidney capsule implantation and injection of CD34+ cells. Within the leukocytes singlet population, we gated cells positive for human CD45 (left panel). Within the CD45+ population, we gated CD19 + B cells and CD3+ T cells (middle panel) as well as CD14+ monocytes (right panel)

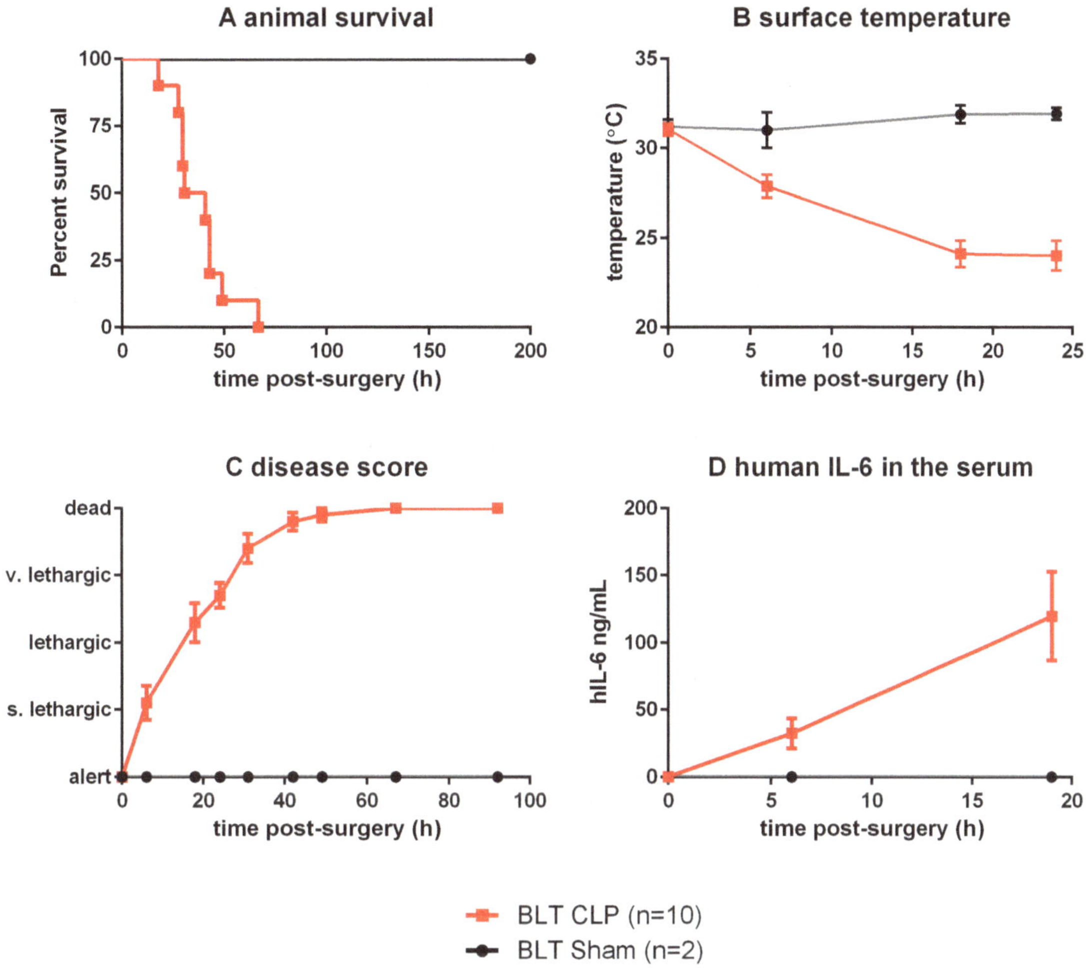

Fig. 3 CLP induces sepsis in BLT humanized mice. BLT mice were subject to CLP to induce sepsis or sham surgery as a control. We monitored (**a**) animal survival, (**b**) surface temperature, showing hypothermia, and (**c**) disease score, for example, degree of lethargy. (**d**) Human IL-6 was measured in serum samples collected presurgery and at 6 and 18 h postsurgery

3.8 Sepsis in BLT Humanized Mice

Once the lymphoid compartment is reconstituted, sepsis can be induced in humanized mice, via cecal ligation and puncture (CLP). Alternatively, the investigator may utilize one of the many other sepsis models described in this book. In Fig. 3, we show results for BLT mice subject to CLP surgery. For CLP, we tied off 1 cm of the cecum and puncture once through and through with a 21G needle (details of our CLP method are described in our prior reports [17–20]). Both CLP and sham mice received 1 mL lactated Ringer's solution postoperatively and antibiotics (imipenem–cilastatin, 25 mg/kg) twice daily for 5 days. Animals subject to CLP showed rapid mortality (Fig. 3a), elevated disease score (degree of lethargy, Fig. 3b) and hypothermia (Fig. 3c), which were absent in mice subject to sham surgery. We also detected high levels of human

IL-6 in the sera of animals subject to CLP, but not those subject to sham surgery (Fig. 3d). Interestingly, in our hands the humanized mice showed 100% mortality despite antibiotics and fluid resuscitation suggesting that they are more highly susceptible to sepsis than the C57BL/6 mice that we have previously employed [17–20]. Please note that this study was approved by the IACUC of TTUHSC at El Paso.

4 Notes

1. The sex of the mice will affect reconstitution efficiency. Female mice show improved reconstitution relative to males [21]. However, both genders may be employed for scientific rigor. Younger mice (6 weeks) may demonstrate the best reconstitution.
2. Liver and thymus samples should be procured from second trimester (17–20 week) fetal cadavers. The material should be used on the day of receipt. Please note that this material poses a risk of exposure to human pathogens. Contact your institution's biosafety office to determine the appropriate biosafety level and handing conditions. Please check local regulations governing the transfer and use of fetal tissues.
3. This protocol uses a manual magnet separator (MidiMACS or QuadroMACS). However, an automated MACS Cell Separation Instrument can also be used. Follow the manufacturer's instructions for this instrument. Other isolation kits could also be used, such as the EasySep Human CD34 Positive Selection Kit II (Stem Cell Technologies) or the Dynabeads CD34 Positive Isolation Kit (Thermo Fisher Scientific).
4. The purity of CD34+ cells can be determined by antibody staining and flow cytometry, using the protocol described in Subheading 3.7, from **step 6** onward, substituting a fluorescently labeled antibody against human CD34 in **step 8**.
5. Thermal support should be provided during surgery and postoperative recovery. Appropriate options include a recirculating water heater pad, an infrared heating pad or a microwaveable heating pad. For the latter device, take care not to overheat in the microwave. Typically, separate heating pads are allocated for intraoperative and postoperative use.
6. Press-and-seal food wrap has been shown to be reasonably sterile and makes an excellent surgical drape for mice, covering and immobilizing the animals during surgery. Alternatively, a square of standard surgical drape may be used.
7. Surgical implements should be cleaned with ethanol and sterilized by autoclave prior to the starting surgery on the first

mouse. Instruments can be cleaned with ethanol pads and resterilized in a benchtop bead sterilizer prior reuse on additional mice. The fine-tipped Dumont tweezers will dull quickly and need to be replaced at frequent intervals.

8. Tail vein injection can be performed with the mouse sedated under isoflurane anesthesia, or with the mouse conscious and secured in a restraint device.
9. Fc blocking reagent will reduce nonspecific binding of antibodies to macrophages. We have found that there is not a great deal of nonspecific staining in the absence of Fc block for this protocol, however this may vary depending on the antibodies used.
10. When overlaying the cell suspension, hold the tube at an angle and apply the pipette tip to the lower wall of the tube, just at the surface of the Ficoll-Paque liquid. Slowly release the cell suspension from the pipette to create a separate layer. Carefully extract the pipette and return the tube to the vertical position.
11. The expected yield of cells from the liver specimen is approximately 3–6 $\times$ 10^8, and the expected yield of enriched CD34+ cells is approximately 1–2 $\times$ 10^7. Each LS column is capable of handling a maximum starting number of 1 $\times$ 10^9 cells, and enriching up to 10^8 magnetically labeled cells from this mixture. Therefore one LS column will typically suffice per specimen.
12. This protocol follows the manufacturer's instructions for CD34+ enrichment, however the investigator is advised to check their kit's documents for updates to the protocol.
13. Contact your institution's biosafety officer to determine your institution's regulations and appropriate practices for the use of a fixed radioactive source. Badge dosimetry and radiation dose meters may be employed to monitor exposure.
14. Some irradiators require that animals are placed in an Allentown-style cage, while others require that animals are placed in a pie-type restraint device. Check the instructions for your irradiator.
15. Once mice are irradiated, they will be severely immunodeficient. They must be administered CD34+ cells within 1 day to begin the reconstitution process, otherwise they will be highly susceptible to infection by environmental microbes. Kidney capsule and intravenous injection of CD34+ cells must be performed *after* the irradiation to avoid killing the leukocytes in these preparations.
16. This protocol uses isoflurane anesthesia. If the investigator does not have this apparatus, other anesthetic agents may be used such as ketamine–xylazine. Ketamine is a controlled

substance and may require a special license. Please contact your institutional veterinarian for advice on appropriate anesthetic regimens and your local regulatory authorities to determine the licensing requirements for controlled substances.

17. During surgery, the mouse should be sedated under Stage III, plane 2 anesthesia. At this stage, the animal's respiratory rate is decreased and may appear slightly shallow but not gasping. The heart rate is decreased but not markedly decreased, and the animal is unresponsive to toe pinch and noxious stimuli. Signs of responsiveness include pedal reflex (pulling back the leg) after toe pinch, physical movement in response to noxious stimulus (such as jerking when the skin is cut), as well as more subtle signs such as changes in the breathing pattern or heart rate. Animals that do not exhibit adequate anesthetic depth should be given a higher rate of isoflurane (the machine isoflurane setting should be increased in gradual increments).
18. This protocol uses Buprenorphine SR (sustained release) as this provides continuous pain relief appropriate for this major survival surgery, with fewer injections than standard Buprenorphine. Other analgesics may be suitable; contact your institutional veterinarian for advice. Buprenorphine is a controlled substance and may require a license. Please contact your local regulatory authorities to determine the requirements for your region.
19. Clean the clippers with the clipper disinfectant and lubricant spray at the start of each set of surgeries and any time the clippers become clogged (indicated by difficulty in shaving). It can help to run a thin piece of paper towel in between the blades to remove fur and excess disinfectant and lubricant spray. It is important to be careful when shaving because the mouse skin is delicate and may become torn. Shaving is facilitated by holding the mouse in a modified scruff position to pull the skin taught over the area that is being shaved.
20. Postoperative complications are rare but infection of the wound or abdominal cavity can occur. To avoid this, antibiotics should be provided postoperatively. The kidney capsule implantation surgery is expected to result in a mild to moderate degree of pain for 1–2 days after surgery, indicated by lethargy, scruffiness, and a hunched posture: these symptoms are alleviated through the use of Buprenorphine, as described above. If a more substantial, unanticipated degree of pain is observed, as indicated by a worsening in the symptoms described above, or if the animals develops signs of wound or abdominal cavity infection (indicated by redness and/or swelling of the abdomen, and/or seeping of fluid from the surgical wound) the institutional veterinarian should be contacted for assessment of the animal.

21. Alternately, sulfatrim oral antibiotic suspension (6.75 mL in 200 mL) may be administered in the drinking water (your institutional veterinarian can provide advice).
22. When using a restraint device, the animal should be secured tightly enough so that it cannot move, but not so tight that it impedes its breathing. Check the mouse regularly to ensure its welfare.
23. We typically inject 5×10^5 cells to achieve maximum reconstitution. However the procedure can be performed with as few as 1×10^5 cells. In this case, adjust the concentration of the cell suspension to 5×10^5 cells/mL by adding an appropriate volume of PBS, before injecting 200 μL volume per mouse.
24. When the needle is in the correct position, the plunger should move smoothly. If the plunger is difficult to depress it is likely not in the vein. Remove the needle and try again. With repeat attempts at tail vein injection, one should move from a distal site to a proximal site on the tail vein (closer to the body), otherwise the injected solution will come out of the more proximal hole. The opposite tail vein may also be used if multiple attempts are required. Adjusting the height of the platform and the room lighting may also help to better visualize and inject the vein. With practice, it should be possible to inject the tail vein with one or two attempts.
25. It may not be possible to visualize a cloudy band at the interface of the Histopaque-1077, or a pellet in the subsequent steps, due to the low number of lymphocytes that will be present in this small blood sample. Collect material from the interface region and proceed through the protocol, whether or not cells can be visualized.
26. Cells are typically stained in final volume of 50 or 100 μL. The volume of FACS staining buffer required is equal to the final volume minus the amount of antibody and Fc block that will be added. For example, if the cells will be stained in a final volume of 100 μL and 2.5 μL FC blocking reagent will be added plus 20 μL antibodies, then the cells should be resuspended in 77.5 mL FACS staining buffer. Refer to the product information sheets for the antibodies that you have purchased, and the Fc blocking reagent, if used, to determine appropriate volumes. A titration should be performed to determine the optimal concentration of each antibody for staining. Human blood can be used for this titration curve.

Acknowledgments

This work was supported by a seed grant from TTUHSC El Paso. We would like to thank Dinesh Goswami and Blake Mireles for assistance with photography.

References

1. Lagu T, Rothberg MB, Shieh MS et al (2012) Hospitalizations, costs, and outcomes of severe sepsis in the United States 2003 to 2007. Crit Care Med 40:754–761
2. Angus DC, Linde-Zwirble WT, Lidicker J et al (2001) Epidemiology of severe sepsis in the United States: analysis of incidence, outcome, and associated costs of care. Crit Care Med 29:1303–1310
3. Takao K, Miyakawa T (2015) Genomic responses in mouse models greatly mimic human inflammatory diseases. Proc Natl Acad Sci U S A 112:1167–1172
4. Seok J, Warren HS, Cuenca AG et al (2013) Genomic responses in mouse models poorly mimic human inflammatory diseases. Proc Natl Acad Sci U S A 110:3507–3512
5. Osuchowski MF, Remick DG, Lederer JA et al (2014) Abandon the mouse research ship? Not just yet! Shock 41:463–475
6. Remick D (2013) Use of animal models for the study of human disease-a shock society debate. Shock 40:345–346
7. Smith DJ, Lin LJ, Moon H et al (2016) Propagating humanized BLT mice for the study of human immunology and immunotherapy. Stem Cells Dev 25:1863–1873
8. Melkus MW, Estes JD, Padgett-Thomas A et al (2006) Humanized mice mount specific adaptive and innate immune responses to EBV and TSST-1. Nat Med 12:1316–1322
9. Lan P, Tonomura N, Shimizu A et al (2006) Reconstitution of a functional human immune system in immunodeficient mice through combined human fetal thymus/liver and CD34+ cell transplantation. Blood 108:487–492
10. Ye C, Choi JG, Abraham S et al (2012) Human macrophage and dendritic cell-specific silencing of high-mobility group protein B1 ameliorates sepsis in a humanized mouse model. Proc Natl Acad Sci U S A 109:21052–21057
11. Unsinger J, McDonough JS, Shultz LD et al (2009) Sepsis-induced human lymphocyte apoptosis and cytokine production in "humanized" mice. J Leukoc Biol 86:219–227
12. Skirecki T, Kawiak J, Machaj E et al (2015) Early severe impairment of hematopoietic stem and progenitor cells from the bone marrow caused by CLP sepsis and endotoxemia in a humanized mice model. Stem Cell Res Ther 6:142
13. Skirecki T, Drechsler S, Hoser G et al (2019) The fluctuations of leukocytes and circulating cytokines in septic humanized mice vary with outcome. Front Immunol 10:1427
14. Lapko N, Zawadka M, Polosak J et al (2017) Long-term monocyte dysfunction after sepsis in humanized mice is related to persisted activation of macrophage-colony stimulation factor (M-CSF) and demethylation of PU. 1, and it can be reversed by blocking M-CSF in vitro or by transplanting naive autologous stem cells in vivo. Front Immunol 8:401
15. Laudanski K, Lapko N, Zawadka M et al (2017) The clinical and immunological performance of 28 days survival model of cecal ligation and puncture in humanized mice. PLoS One 12:e0180377
16. Walker WE (2018) Methods to study the innate immune response to sepsis. Methods Mol Biol 1717:189–206
17. Goswami DG, Garcia LF, Dodoo C et al (2021) Evaluating the timeliness and specificity of CD69, CD64 and CD25 as biomarkers of sepsis in MICE. Shock 55:507–518
18. Walker WE, Bozzi AT, Goldstein DR (2012) IRF3 contributes to sepsis pathogenesis in the mouse cecal ligation and puncture model. J Leukoc Biol 92:1261–1268
19. Heipertz EL, Harper J, Walker WE (2017) STING and TRIF contribute to mouse sepsis, depending on severity of the disease model. Shock 47:621–631
20. Heipertz EL, Harper J, Lopez CA et al (2018) Circadian rhythms influence the severity of sepsis in mice via a TLR2-dependent, leukocyte-intrinsic mechanism. J Immunol 201 (1):193–201
21. Notta F, Doulatov S, Dick JE (2010) Engraftment of human hematopoietic stem cells is more efficient in female NOD/SCID/IL-2Rgc-null recipients. Blood, the journal of the American society of. Hematology 115:3704–3707

Chapter 13

Scoring Sepsis Severity in Mice

Tina S. Mele

Abstract

A reliable scoring system that predicts the development of sepsis, septic shock, and death enables comparison of disease severity and treatment outcomes in animal models of sepsis. Mice are used in the majority of preclinical sepsis studies. We describe a murine sepsis score that evaluates seven clinical variables in an experimental mouse model of polymicrobial sepsis.

Key words Sepsis severity score, Mouse model of sepsis

1 Introduction

Sepsis is the leading cause of death in hospitalized patients with a mortality rate of approximately 30% despite optimal treatment [1]. To gain further insights into the pathophysiology of sepsis and to evaluate new treatments prior to testing in clinical trials, most preclinical sepsis studies utilize mouse models of sepsis. Thus, a reliable, sensitive, and specific scoring system is required to compare disease outcomes, evaluate therapeutic regimens and investigate novel diagnostic biomarkers [2–4]. We describe a validated murine sepsis score (MSS) that can consistently predict the development of sepsis, septic shock, and death [5]. The MSS has several important advantages. The MSS is based on simple observations; thus, it may be easily implemented in any laboratory that utilizes a mouse model of sepsis. Furthermore, the MSS can be applied in sepsis experiments with any investigative timeline because of its noninvasive nature.

2 Materials

2.1 Animals and Experimental Supplies

1. Mice subject to an experimental sepsis model (*see* **Notes 1** and **2**).
2. Mouse cages with removable lids.

Wendy E. Walker (ed.), *Sepsis: Methods and Protocols*, Methods in Molecular Biology, vol. 2321,
https://doi.org/10.1007/978-1-0716-1488-4_13,

3. Food and water, supplied as outlined by experimental conditions.
4. Buprenorphine (0.1 mg/kg) for analgesia.
5. 27G needles and 3-mL syringes for buprenorphine injection.

2.2 Sepsis Scoring

1. Paper or electronic copy of scoring sheets (Table 1).
2. Electronic timer.

3 Methods

The murine sepsis score (MSS) was validated in a mouse model of polymicrobial intra-abdominal sepsis. Approximately 20% of patients hospitalized with sepsis have an abdominal source of infection [6]. The MSS scoring system uses easily observed mouse behaviors and changes in appearance that have been identified as effective measures of septic morbidity [7]. The MSS evaluates seven individual criteria and results in an aggregate score to evaluate the clinical condition of mice with experimental sepsis. Disease severity and risk of ensuing mortality are monitored by the MSS at appropriate intervals during the experimental timeline. Depending on the microbial load administered, or the sepsis model utilized, the timeline for mouse deterioration will vary.

3.1 Sepsis Severity Monitoring

1. Mice should be given ad libitum access to food and water throughout the experimental timeline and housed in social groups of 1–3 cage mates (*see* **Note 3**).
2. Induce sepsis as specified in the relevant experimental sepsis protocol. Validation of the MSS was conducted in a fecal-induced peritonitis (FIP) model (*see* **Note 2**).
3. Assess pain using mouse facial expression, body posture, vocalization, and behavior (*see* **Note 4**).
4. Provide analgesia by subcutaneous injection of buprenorphine (0.1 mg/kg).
5. Seven variables are assessed to obtain the MSS including spontaneous activity, response to touch and auditory stimuli, posture, respiration rate and quality (labored breathing or gasping), and appearance (i.e., degree of piloerection). Each of the seven variables is given a score between 0 and 4 (Table 1).
6. Mice are assessed every 2 h after the induction of sepsis for 12 h, and then every hour thereafter (*see* **Note 5**). Mice are evaluated while in their cages with the lids removed for better visualization and assessment.

Table 1
Murine Sepsis Score (MSS) to assess the severity of disease in an experimental model of fecal-induced peritonitis

Variable	Score and description
Appearance	0. Coat is smooth 1. Patches of hair piloerected 2. Majority of back is piloerected 3. Piloerection may or may not be present, mouse appears “puffy” 4. Piloerection may or may not be present, mouse appears emaciated
Level of consciousness	0. Mouse is active 1. Mouse is active but avoids standing upright 2. Mouse activity is noticeably slowed. The mouse is still ambulant 3. Activity is impaired. Mouse only moves when provoked, movements have a tremor 4. Activity severely impaired. Mouse remains stationary when provoked, with possible tremor
Activity	0. Normal amount of activity. Mouse is involved in any of the following activities: Eating, drinking, climbing, running, fighting 1. Slightly suppressed activity. Mouse is moving around bottom of cage 2. Suppressed activity. Mouse is stationary with occasional investigative movements 3. No activity. Mouse is stationary 4. No activity. Mouse experiencing tremors, particularly in the hind legs
Response to stimulus	0. Mouse responds immediately to auditory stimulus or touch 1. Slow or no response to auditory stimulus, strong response to touch (moves to escape) 2. No response to auditory stimulus, moderate response to touch (moves a few steps) 3. No response to auditory stimulus, mild response to touch (no locomotion) 4. No response to auditory stimulus. Little or no response to touch. Cannot right itself if pushed over
Eyes	0. Open 1. Eyes not fully open, possibly with secretions 2. Eyes at least half closed, possibly with secretions 3. Eyes half closed or more, possibly with secretions 4. Eyes closed or milky
Respiration rate	0. Normal, rapid mouse respiration 1. Slightly decreased respiration (rate not quantifiable by eye) 2. Moderately reduced respiration (rate at the upper range of quantifying by eye) 3. Severely reduced respiration (rate easily countable by eye, 0.5 s between breaths) 4. Extremely reduced respiration (>1 s between breaths)
Respiration quality	0. Normal 1. Brief periods of labored breathing 2. Labored breathing, no gasping 3. Labored with intermittent gasping 4. Gasping

7. Mice are euthanized if the MSS at any time point is greater than 21, or if the points attributed to respiratory rate or quality have increased by more than 3 points.

3.2 Sepsis Scoring and Prediction of Mortality

1. In the FIP model, using the 90 mg/mL fecal slurry concentrations, the MSS predicts a mortality rate of 40% within 1 h and 75% within 2 h when the MSS is 10 or greater. Euthanasia should be considered based on the condition of the mouse.
2. Once a mouse develops a score of 3 or greater, the MSS has a 100% specificity for predicting onset of developing septic shock and death within 24 h.

4 Notes

1. In our experiments we utilize mice 10 to 12 weeks of age (weight 20–25 g before surgery).
2. Validation of the MSS was conducted in an FIP sepsis model. The MSS may also be used in other models of sepsis, including microbe-specific models, although calibration of the scoring system will be necessary to optimize sensitivity and specificity. In our FIP model, sepsis was induced with the following protocol: Feces are collected from the lower cecum of euthanized donor mice, weighed and mixed with sterile normal saline to obtain the desired fecal concentration. Each mouse is given a 0.5 mL intraperitoneal (i.p.) injection of the prepared fecal slurry solution using a syringe and 27G needle. Depending on the fecal slurry (FS) concentrations used (45, 90, and 180 mg/mL), FIP mice have a survival rate of 0% for 180 mg/mL FS ($n = 20$), 25% for 90 mg/mL FS ($n = 200$) and 40% for 45 mg/mL FS ($n = 20$) at 24 h post-FIP induction [5].
3. Depending on the illness severity, we isolate mice to prevent cannibalism during the experiment as mice can exhibit this behavior during stress.
4. Facial assessment of pain was based on observations as described by Langford et al. [8] which monitors five facial features including orbital tightening, nose bulge, check bulge, ear position and whisker changes to determine if the mouse is experiencing pain.
5. The frequency of assessment will vary with the severity of sepsis induced by the specific sepsis experimental model utilized. In our FIP model, we monitor mice as described in the protocol. Between 0 and 11 h post-FIP induction, mouse scores remain relatively consistent. After 12 h post-FIP induction, septic mice begin to manifest impaired respiratory status due to decreased respiratory rate and increasingly labored breathing with minimal response to auditory and tactile stimuli.

References

1. Martin GS, Mannino DM, Eaton S et al (2003) The epidemiology of sepsis in the United States from 1979 through 2000. N Engl J Med 348:1546–1554
2. Buras JA, Holzmann B, Sitkovsky M (2005) Animal models of sepsis: setting the stage. Nat Rev Drug Discov 4:854–865
3. Esmon CT (2004) Why do animal models (sometimes) fail to mimic human sepsis? Crit Care Med 32:S219–S222
4. Marshall JC, Reinhart K, International Sepsis F (2009) Biomarkers of sepsis. Crit Care Med 37:2290–2298
5. Shrum B, Anantha RV, Xu SX et al (2014) A robust scoring system to evaluate sepsis severity in an animal model. BMC Res Notes 7:233
6. Vincent JL, Rello J, Marshall J et al (2009) International study of the prevalence and outcomes of infection in intensive care units. JAMA 302:2323–2329
7. Nemzek JA, Hugunin KM, Opp MR (2008) Modeling sepsis in the laboratory: merging sound science with animal well-being. Comp Med 58:120–128
8. Langford DJ, Bailey AL, Chanda ML et al (2010) Coding of facial expressions of pain in the laboratory mouse. Nat Methods 7:447–449

Chapter 14

Identification of ILC2 in the Lung Using Flow Cytometry

Hui Xu and Meihong Deng

Abstract

Group 2 innate lymphoid cells (ILC2), despite their scarcity, are the dominant innate lymphoid cell population in the lung, orchestrating innate immunity and adaptive immunity (Germain and Huang. Curr Opin Immunol 56:76–81, 2019; Krabbendam et al. Immunol Rev 286:74–85, 2018; Mindt et al. Front Immunol 9:840, 2018). Recent studies reveal that ILC2 play critical roles in inflammation-associated lung injury during sepsis (Lai et al. Cell Death Dis 9:369, 2018; Xu et al. Immunol Cell Biol 96:935–947). Therefore, studies aiming to understand the pathobiology of ILC2 may reveal new therapeutic strategies for sepsis. However, the identification of ILC2 requires multiple surface and intracellular markers. This makes the detection of ILC2 in the lung challenging. Here we describe a method to detect ILC2 in the mouse lung using flow cytometry.

Key words ILC2, Lung, Flow cytometry

1 Introduction

The Innate lymphoid cell (ILC) family includes the three different subsets of noncytotoxic ILCs (ILC1, ILC2, and ILC3) [6]. ILCs have recently been recognized as important immunoregulators in inflammatory diseases [1, 2]. ILC2, as a subpopulation of ILCs, play important roles in infection, inflammation, and allergic diseases [3, 4]. As a subpopulation of lymphoid cells, ILCs share the classic morphology with other lymphoid cells. Distinguishing them from other lymphoid cells, ILCs do not express the major surface molecules present on other immune cell types, such as CD3, CD19, CD56, CD68, and CD205; and are therefore designated as cell lineage marker negative (Lin-) [6]. Furthermore, they do not have any antigen specificity for the recognition of pathogens [7]. ILC2 are characterized as Lin-, CD117+, CD90+, ST2+ in mice [8]. As multiple markers are required for the identification of ILC2, flow cytometry is used widely for their detection [5, 9].

Wendy E. Walker (ed.), *Sepsis: Methods and Protocols*, Methods in Molecular Biology, vol. 2321,
https://doi.org/10.1007/978-1-0716-1488-4_14, © Springer Science+Business Media, LLC, part of Springer Nature 2021

Flow cytometry is a well-established laser-based technique for phenotyping cells in a single-cell suspension. Fluorescently labeled cells are run through a laser beam so that light is first absorbed and then emitted in a band of wavelengths. Flow cytometry allows for simultaneous multiparametric quantitative analysis of the physical and chemical characteristics of cells in suspension. This makes flow cytometry ideal for the detection of ILC2.

In this chapter, we describe the method of detecting ILC2 in the lung using flow cytometry. We provide step-by-step instructions including detailed procedures for lung tissue collection, preparation of a single-cell suspension, flow cytometry, and data analysis.

2 Materials

2.1 Lung Tissue Collection

1. Male or female C57BL6 mice, 8–10 weeks old.
2. Surgical instrument: scissors, needle holder, forceps.
3. PBS: phosphate-buffered saline pH 7.2.

2.2 Single-Cell Suspension Preparation

1. PBS: phosphate-buffered saline pH 7.2. Keep buffer cold (2 – 8 °C).
2. PEB buffer: Prepare a solution containing PBS, pH 7.2, 0.5% bovine serum albumin (BSA), and 2 mM EDTA. Always use freshly prepared buffer. Keep buffer cold (2–8 °C).
3. GentleMACS C Tubes (Miltenyi Biotec).
4. Lung Dissociation Kit, mouse (Miltenyi Biotec).
5. Scissors.
6. Tube rotator.
7. 37 °C incubator.
8. GentleMACS Dissociator (Miltenyi Biotec).
9. Cell strainer-70 μm.
10. 1× RBC Lysis Buffer (Thermo Fisher).
11. Centrifuge.

2.3 Flow Cytometry

1. PBS: phosphate-buffered saline pH 7.2. Keep buffer cold (2–8 °C).
2. Protein transport inhibitor cocktail (containing Brefeldin A and Monensin).
3. Phorbol myristate acetate (PMA).
4. Ionomycin.
5. Purified anti-mouse CD16/32 Antibody (FC block).
6. Fluorochrome-conjugated antibodies (Table 1).

Table 1
Antibodies used in flow cytometry

Name
Fixable viability dye eFluor® 506
Anti-mouse CD45 Alexa Fluor® 700
Anti-mouse CD3e FITC
Anti-mouse CD11b FITC
Anti-mouse CD127 eFluor® 450
Anti-mouse Ly-6A/E (Sca-1) PE
Anti-mouse CD90.2 APC eFluor 780
Anti-mouse CD11c APC-eFluor® 780
Anti-mouse CD117 (c-Kit) PE-Cyanine7
Anti-mouse CD25 PE
Anti-mouse Ly-6G (Gr-1) PE-eFluor® 610
Anti-mouse F4/80 PE
Anti-mouse CD11b eFluor® 450
Anti-mouse ST2 PerCP-eFluor® 710
Anti-mouse GATA3 PE-CF594
Mouse IgG1 κ isotype control PE-CF594
Anti-mouse F4/80 APC/Cy7

7. 4% Paraformaldehyde in PBS.
8. Intracellular Fixation & Permeabilization Buffer Set (eBioscience).
9. Flow cytometer, capable of reading 10 or more fluorochromes.

2.4 Data Analysis

1. Flow cytometry data analysis software.

3 Methods

3.1 Lung Tissue Collection

1. Euthanize the mice with CO_2 and withdraw blood by cardiac puncture.
2. Perfuse the lung lobes with sterile PBS via the right ventricle to remove erythrocytes (as described in a prior report [10]).
3. Remove the lung and place the tissue into gentleMACS C tubes and store on ice.

3.2 Single-Cell Suspension Preparation

The lung tissue is dissociated using the Lung Dissociation Kit, mouse (Miltenyi Biotec) according to the manufacturer's protocol.

1. Prepare the enzyme mix by adding 2.4 mL of 1× Buffer S, 100 μL of Enzyme D, and 15 μL of Enzyme A (Lung Dissociation Kit) into a gentleMACS C Tube.
2. Place the lung tissue in the tube with the enzyme mix. Mince the lung tissue with scissors.
3. Tightly close the C Tube and attach it onto the sleeve of the gentleMACS Dissociator, following the manufacturer's instructions.
4. Run the gentleMACS Program m_lung_01.
5. After termination of the program, detach the C Tube from the gentleMACS Dissociator and incubate the sample for 30 min at 37 °C, under continuous rotation.
6. After incubation, attach the C tube onto the gentleMACS Dissociator and run the gentleMACS Program m_lung_02, following the manufacturer's instructions.
7. After termination of the program, resuspend the sample in 10 mL PEB buffer and apply the cell suspension to a cell strainer (70 μm) placed on top of a 5 mL tube (*see* **Note 1**).
8. Wash the cell strainer twice with 2.5 mL PEB buffer.
9. Discard the cell strainer and centrifuge the cell suspension at 300 × *g* for 10 min. Aspirate the supernatant completely.
10. (Optional) Resuspend the cells in 2 mL 1× RBC lysis buffer and incubate at room temperature for 5 min. Terminate the reaction by adding 10 mL PEB buffer and centrifuge the cell suspension at 300 × *g* for 10 min. Aspirate the supernatant completely.
11. Resuspend the cells in 100 μL PEB buffer and process the cells immediately for antibody staining.

3.3 Antibody Staining and Flow Cytometry

1. To block secretion of GATA-3, stimulate cells with PMA (10 ng/mL), protein transport inhibitor containing Brefeldin A and Monensin (1 μL/mL) and ionomycin (500 ng/mL) for 3 h at 37 °C. Stimulation with PMA and ionomycin activates cells and protein transport inhibitor prevents protein transport to the extracellular space (*see* **Note 2**).
2. For setting compensation, prepare a tube of unstained cells, and a single antibody stain for each fluorochrome used in the experiment. The experimental samples will be stained with all of the antibodies (Table 1), to detect ILC2.
3. Wash the cells with 2.5 mL PBS buffer and centrifuge the cell suspension at 500 × *g* for 5 min.

4. Aspirate the supernatant and resuspend the cells in 100 μL PBS buffer.
5. Incubate the cells with fixable viability dye for 20 min at 4 °C in the dark (*see* **Note 3**).
6. Wash the cells with 2.5 mL PBS buffer and centrifuge the cell suspension at 500 × *g* for 5 min.
7. Aspirate the supernatant and resuspend the cells in 100 μL PBS buffer.
8. Incubate the cells with purified anti-mouse CD16/CD32 (1 μL/tube) at 4 °C for 10 min to block Fc receptors (*see* **Note 4**).
9. Incubate the cells with a combination of fluorochrome-conjugated antibodies including anti-mouse CD45, lineage markers (Lin, including CD11b, CD3e, CD45R, NK1.1), anti-mouse CD90.2, anti-mouse ST2, anti-mouse sca-1, anti-mouse CD117(c-Kit), anti-mouse CD127 and anti-mouse CD25 for 30 min at 4 °C in the dark (*see* **Note 5**).
10. Wash the cells with 2.5 mL PBS buffer and centrifuge the cell suspension at 500 × *g* for 5 min.
11. Aspirate the supernatant and resuspend the cells in 100 μL PBS buffer.
12. Using the intracellular fixation and permeabilization buffer set, incubate the cells with 100 μL IC fixation buffer for 30 min at room temperature in the dark.
13. Using the intracellular fixation and permeabilization buffer set, wash the cells with 2 mL 1× permeabilization buffer and centrifuge the cell suspension at 500 × *g* for 5 min.
14. Aspirate the supernatant and resuspend the cells in 100 μL 1× permeabilization buffer.
15. Wash the cells with 2 mL 1× permeabilization buffer and centrifuge the cell suspension at 500 × *g* for 5 min.
16. Aspirate the supernatant and resuspend the cells in 100 μL 1× permeabilization buffer.
17. Incubate the cells with fluorochrome-conjugated anti-mouse GATA-3 or isotype-matched control antibodies for 40 min at room temperature in the dark.
18. Wash the cells with 2 mL 1× permeabilization buffer and centrifuge the cell suspension at 500 × *g* for 5 min.
19. Aspirate the supernatant, resuspend the cells in 100 μL 1× permeabilization buffer and process the cells immediately for data acquisition.

20. Run the unstained and single stained samples through the flow cytometer and adjust compensation manually or automatically (*see* **Note 6**).
21. Acquired data from the experimental samples using the flow cytometer.

3.4 Data Analysis

1. Analyze the data with flow cytometry analysis software.
2. To identify specific myeloid cell subsets in the lungs, start by gating leukocytes according to their FSC-A: SSC-A profile (Fig. 1).
3. Subsequently, gate singlet cells based on their FSC-H:FSC:W profile (Fig. 1).
4. Next, gate live cells. This is accomplished by excluding cells that take up the viability stain (Fig. 1).
5. Subsequently, gate cells with CD45 expression (Fig. 1).
6. To further phenotype ILC2, compare the expression of Lin, CD90, GATA-3, CD117, ST2, CD127, CD25, and Sca-1. ILC2 are identified as $CD45^+/Lin^-/CD90.2^+/ST2^+/sca\text{-}1^+/GATA\text{-}3^+/CD117^+/CD127^+/CD25^+$ (Fig. 1).

4 Notes

1. The lung lobes may not be completely dissociated after this step. In this case, incubate the sample for an additional 15 min at 37 °C under continuous rotation. After incubation, attach the C tube onto the gentleMACS Dissociator and run the gentleMACS Program m_lung_02.
2. Addition of the protein transport inhibitor cocktail (containing Brefeldin A and Monensin) to cells in culture results in the accumulation of secreted proteins within cells, and this accumulation can then be detected by intracellular staining and flow cytometric analysis.
3. A viability stain can be utilized to exclude dead cells from the analysis. Ensure you select a viability dye that is suitable to be read after fixation, such as Fixable Viability Dye eFluor® 506.
4. Fc receptors are present on macrophages, dendritic cells and B cells, and can bind nonspecifically to the fluorochrome-conjugated antibodies. The addition of anti-mouse CD16/CD32 prevents this nonspecific binding.
5. Each antibody should be titered to determine the optimal concentration for flow cytometry. Table 1 provides details for a panel of antibodies that produce good results in identifying ILC2, but other fluorochromes may be used. When selecting fluorochromes, make note of which ones work well together on

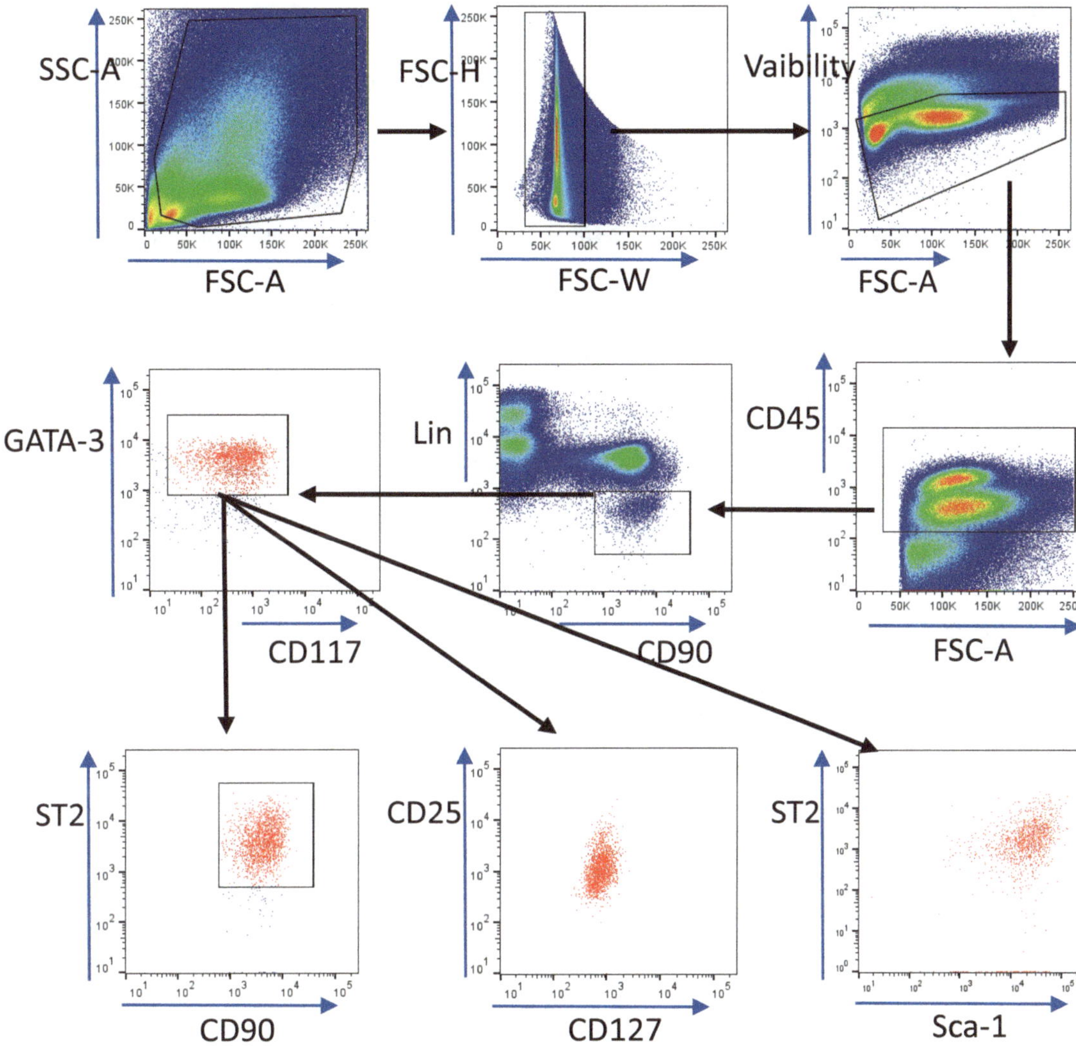

Fig. 1 Flow cytometry plots showing gating of ILC2 in the lungs. After gating leukocytes and singlets, the viable CD45+/Lin-/CD90.2+ subset was gated and analyzed for expression of GATA-3, CD117, ST2, CD127, CD25, and sca-1. ILC2 were identified as CD45+/Lin-/CD90.2+/ST2+/sca-1+/GATA-3+/CD117+/CD127+/CD25 cells

the flow cytometer that will be used to read the samples (details can be obtained from the instrument manufacturer). Antigens that are weakly expressed should ideally be conjugated to fluorochromes with a strong fluorescence signal.

6. The emissions spectra of fluorochromes sometimes overlap. For most flow cytometers it is therefore necessary to employ a mathematical correction to correct for this signal overlap. This can be done manually or automatically (by the software operating the machine). The manufacturer of your flow cytometer and acquisition software package can provide further details on how to complete this process.

Acknowledgments

This work was supported by NIH grant R01AI152044 (MD).

References

1. Germain RN, Huang Y (2019) ILC2s—resident lymphocytes pre-adapted to a specific tissue or migratory effectors that adapt to where they move? Curr Opin Immunol 56:76–81
2. Krabbendam L, Bal SM, Spits H et al (2018) New insights into the function, development, and plasticity of type 2 innate lymphoid cells. Immunol Rev 286:74–85
3. Mindt BC, Fritz JH, Duerr CU (2018) Group 2 innate lymphoid cells in pulmonary immunity and tissue homeostasis. Front Immunol 9:840
4. Lai D, Tang J, Chen L et al (2018) Group 2 innate lymphoid cells protect lung endothelial cells from pyroptosis in sepsis. Cell Death Dis 9:369
5. Xu H, Xu J, Xu L et al (2018) Interleukin-33 contributes to ILC2 activation and early inflammation-associated lung injury during abdominal sepsis. Immunol Cell Biol 96:935–947
6. Spits H, Artis D, Colonna M et al (2013) Innate lymphoid cells--a proposal for uniform nomenclature. Nat Rev Immunol 13:145–149
7. Spits H, Di Santo JP (2011) The expanding family of innate lymphoid cells: regulators and effectors of immunity and tissue remodeling. Nat Immunol 12:21–27
8. Artis D, Spits H (2015) The biology of innate lymphoid cells. Nature 517:293–301
9. Xu J, Guardado J, Hoffman R et al (2017) IL33-mediated ILC2 activation and neutrophil IL5 production in the lung response after severe trauma: a reverse translation study from a human cohort to a mouse trauma model. PLoS Med 14:e1002365
10. Arlt MJ, Born W, Fuchs B (2012) Improved visualization of lung metastases at single cell resolution in mice by combined in-situ perfusion of lung tissue and X-Gal staining of lacZ-tagged tumor cells. J Vis Exp 66:e4162

Chapter 15

Measurement of Intestinal Permeability During Sepsis

Takehiko Oami and Craig M. Coopersmith

Abstract

Gut barrier function has been hypothesized to play a critical role in the pathophysiology of sepsis. Measuring intestinal permeability allows for a determination of barrier dysfunction under conditions of health and disease. Fluorescence-conjugated dyes such as fluorescein isothiocyanate-4 kDa dextran (FD4) have been commonly used for evaluating hyperpermeability. Here we describe a common method to measure gut permeability in vivo, following gavage with different sized dyes. In addition, we describe an ex vivo everted gut sac model that allows for discrimination of permeability by segmental geographic location along the intestine.

Key words Tight junction, Gut, Intestine, Permeability, Barrier, Sepsis, Critical care, FD4, Everted gut sac

1 Introduction

The gut has been hypothesized to be the "motor" of critical illness, both driving and then propagating dysregulated inflammation, in part via disruption of its barrier [1]. Intestinal hyperpermeability allows for translocation of microbes, microbial products, and other molecules that ordinarily are restricted to the gut lumen [2]. Improving intestinal permeability by either pharmacologic intervention or genetic manipulation improves survival in murine sepsis, suggesting the potential importance of a leaky gut in human sepsis [3, 4]. Accurately assessing gut permeability allows for measurement of barrier function in preclinical models of sepsis.

The gut barrier is maintained in a semipermeable state in part via expression of tight junction proteins that, depending on their expression and function, lead to increases or decreases in permeability [5]. Gut permeability is mediated through three pathways—two tight junction-dependent (pore and leak) and one tight junction-independent (unrestricted). The pore pathway transports small molecules such as water, nutrients, and ions with a maximum size of 6 Å in a charge-dependent manner [6, 7]. The leak pathway

Wendy E. Walker (ed.), *Sepsis: Methods and Protocols*, Methods in Molecular Biology, vol. 2321,
https://doi.org/10.1007/978-1-0716-1488-4_15,

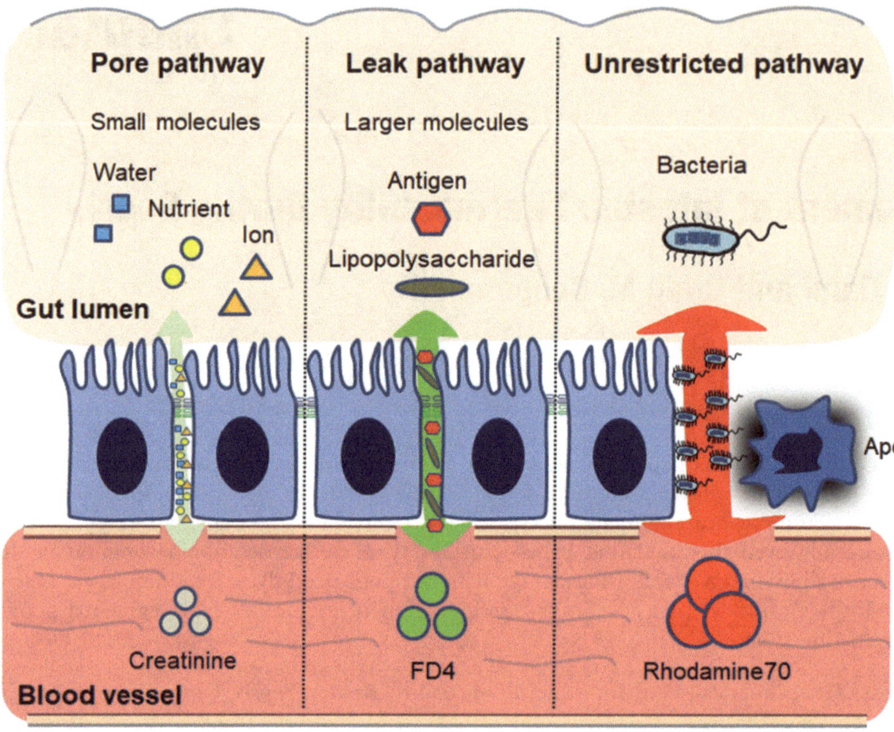

Fig. 1 The pore, leak, and unrestricted pathways are responsible for determining the penetration of substances, depending on the molecular characteristics and pathological condition. The pore pathway (<6 Å) transports small molecules such as water, nutrients, and ions in a charge and size selective manner. The pore pathway can be assayed by gavaging mice with creatinine (6 Å) and measuring its appearance in the bloodstream, Macromolecules such as antigens and lipopolysaccharide nonselectively pass through the leak pathway (<100 Å), which can be assayed with FD4 (28 Å). Unlike the pore and leak pathways which are mediated by tight junctions, the unrestricted pathway is unrelated to tight junctions and is related to apoptosis or epithelial damage, allowing for passage of larger substances including intact bacteria. The unrestricted pathway can be assayed by rhodamine70 (120 Å)

nonselectively transports substances <100 Å such as antigens and lipopolysaccharide [5]. The unrestricted pathway allows free passage of macromolecules and microbes as a result of apoptosis and epithelial damage [8] (Fig. 1).

Oral gavage with fluorescein isothiocyanate-4 kDa dextran (FD4: 28 Å) followed by measuring its appearance in the bloodstream is a commonly used method for assessing gut permeability, dominantly examining the leak pathway. Oral gavage of creatinine (6 Å) and rhodamine B isothiocyanate-70 kDa dextran (rhodamine70: 120 Å) provides additional information regarding the pore and unrestricted pathways [9]. The advantage of this approach is that it allows for an in vivo assessment of permeability. However, this does not allow one to discriminate which portions of the intestine exhibit alterations in permeability. The everted gut sac model is an ex vivo technique that allows for segmental specificity in determining the cause of intestinal hyperpermeability [10, 11]. Here we describe these two complementary methods for assessing gut barrier function.

2 Materials

2.1 Measurement of Intestinal Permeability (In Vivo)

1. Fluorescence-conjugated dyes: FD4 (22 mg/mL) and rhodamine70 (16 mg/mL).
2. Creatinine (40 mg/mL).
3. Feeding needle.
4. Blood collection tubes with ethylenediaminetetraacetic acid (EDTA).
5. 96-well plate.
6. Phosphate-buffered saline (PBS).
7. Creatinine Serum Detection Kit.
8. Fluorescence Plate Reader.

2.2 Everted Gut Sac Model (Ex Vivo)

Overlapping materials from the previous section are not described.

1. Fluorescence-conjugated dye: FD4 (100 μg/mL).
2. Ice-cold Krebs-Henseleit Bicarbonate Buffer (KHBB): pH 7.4, 6.9 g/L sodium chloride, 2.1 g/L sodium bicarbonate, 0.35 g/L potassium chloride, 0.145 g/L magnesium sulfate, 0.145 g/L potassium phosphate monobasic, 0.175 g/L calcium chloride dihydrate, and 2 g/L glucose.
3. Silk braid (4-0).
4. Surgery tools (forceps, scissors, and tweezers).
5. Water bath with temperature control.
6. Oxygen.

3 Methods

3.1 Measurement of Intestinal Permeability (In Vivo)

1. Prepare 22 mg/mL of FD4, 16 mg/mL of rhodamine70, and 40 mg/mL of creatinine in sterile PBS. After preparation, wrap in foil (*see* **Note 1**).
2. Five hours prior to when permeability will be measured, gavage mice with 0.5 mL of creatinine, FD-4, and rhodamine70 dissolved in PBS using a 20-gauge feeding needle (Fig. 2) (*see* **Notes 2–4**).
3. At the time of sacrifice, collect blood in tubes with EDTA and mix well to prevent clots.
4. Spin down the blood at 10,000 rpm for 5 min at 4 °C.

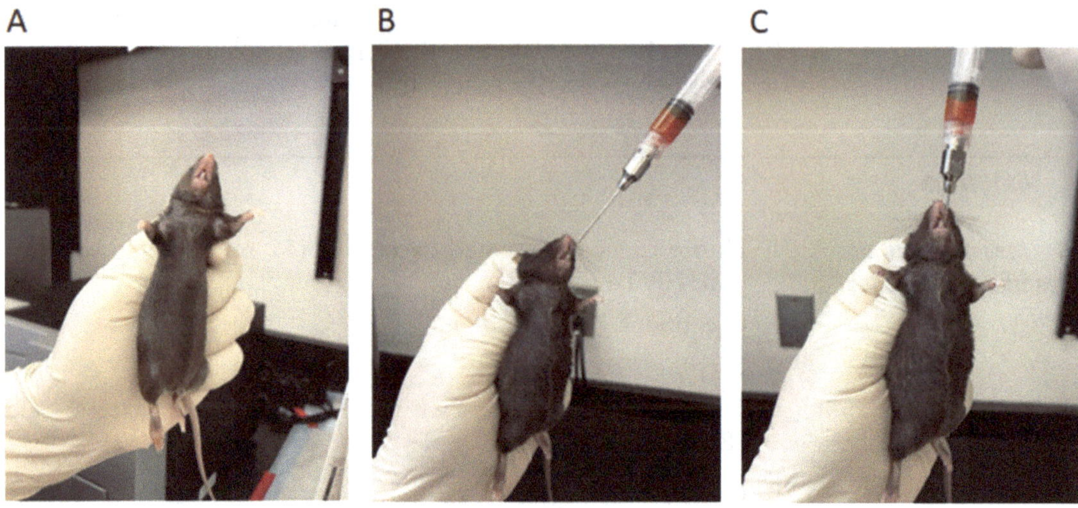

Fig. 2 After appropriate anesthesia, the mouse is held with the neck straight (**a**) so that the feeding needle is easily inserted into the esophagus through the pharynx (**b**) Dyes are then gavaged into the stomach via a feeding needle (**c**)

5. Collect plasma.
6. Dilute plasma 1:2 in PBS (50 μL plasma + 50 μL PBS) and keep on ice.
7. Prepare sequentially diluted dyes as standard.
8. Add 50 μL of diluted plasma samples and standards onto 96 well plates in duplicate.
9. Determine the concentration of FD4 and rhodamine70 using a Fluorescence Plate Reader with an excitation wavelength of 485 and 530 nm and an emission wavelength of 528 and 645 nm.
10. Measure the creatinine concentration with a Creatinine Serum Detection Kit following the manufacturer's protocol (*see* **Note 5**).

3.2 Everted Gut Sac Model (Ex Vivo) (Fig. 3)

1. Prepare 100 μg/mL of FD4 in KHBB (*see* **Note 6**). After preparation, wrap in foil.
2. Remove the gut from the abdomen, flush the intestinal lumen with ice-cold KHBB, and submerge the tissue in the buffer (*see* **Notes** 7 and **8**).
3. Ligate one end of the intestine with 4-0 silk suture.
4. Bring the tip of a feeding needle with a 1 mL syringe close to the knot, holding the ligated intestine. Then, push the knot toward the opposite side to evert the intestine. After flipping over the intestine and keeping the needle inside the sac, fix the position of the everted gut sac over the needle with another silk suture (*see* **Notes 9** and **10**).

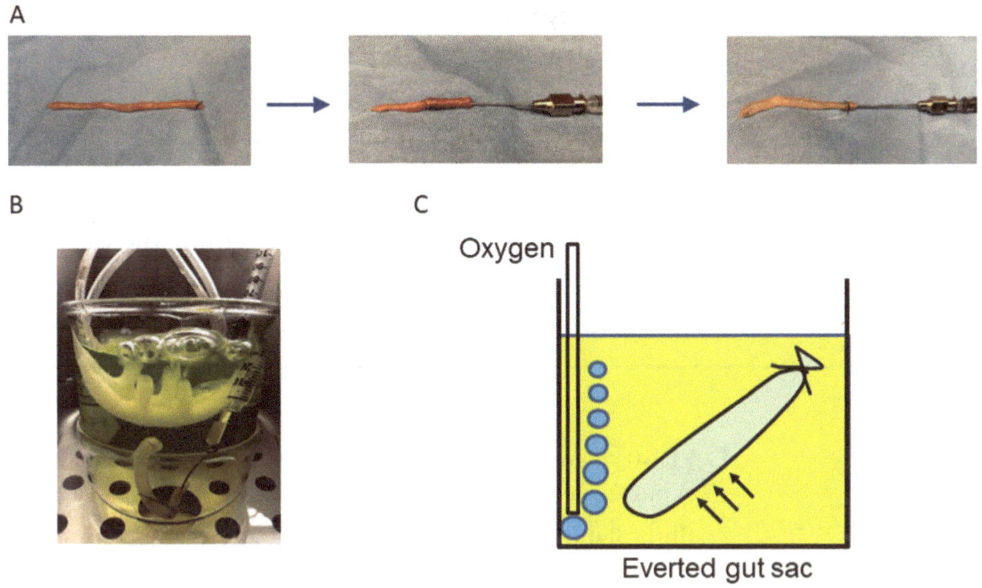

Fig. 3 One end of the segmental jejunum is ligated with 4-0 silk suture (**a**). After being flipped over using the feeding needle, another end of the everted gut sac is ligated so that molecules can be transferred through the mucosa during the incubation. The inflated gut sac is placed in the beaker containing FD4 dissolved in HBSS at 37 °C with 100% oxygen for 30 min (**b**). The schematic (**c**) displays the positional relationship between the gut sac, oxygen, and a water bath. Arrows show the expected direction that FD4 moves through the everted mucosa

5. Inject either 1 mL (small intestine) or 0.5 mL (colon) of KHBB, to inflate the gut sac (*see* **Notes 7–9**).
6. Prepare a water bath at 37 °C and gently aerate with 100% oxygen.
7. Place the everted gut sac in the beaker onto the bath with 100 μg/mL of FD4 in KHBB for 30 min.
8. Collect a 1 mL sample of the buffer from the beaker at the beginning and end of the incubation to determine the initial and final mucosal FD4 concentration, respectively.
9. After 30 min, take the sac out and wash the outside of the gut sac to avoid contamination of beaker fluid on collected samples. After collecting the fluid inside the gut, measure the length and diameter of each sac to standardize the clearance later.
10. Centrifuge the samples at 8,945 × *g* for 5 min and collect the supernatant (*see* **Note 11**).
11. Dilute the fluid 1:2 with PBS, and determine the concentration of FD4 with a Fluorescence Plate Reader, using the method described above (Subheading 3.1, **step 9**).

12. Permeability is calculated as clearance of FD4 from mucosa to serosa in $nL/min/cm^2$ using the following formula:

$$\text{Clearance of FD4} = (\text{serosa FD4 concentration} \times \text{injected volume}) / \text{mucosal FD4 concentration} \times \text{mucosal surface area} \times 30\ \text{min}$$

$$\text{Mucosal surface area} = \pi \times \text{length of the sac} \times \text{diameter of the sac}$$

4 Notes

1. FD40 (molecular weight 40 kDa) may be used instead of FD4. While larger than FD4, FD40 still measures the leak pathway.
2. Blunt handling of the needle during gavage may cause mucosal injury of the digestive tract, leading to absorption directly from vessels. To avoid this, mice should be anesthetized during the procedure. Because deep sedation decreases swallowing, which increases the possibility of regurgitation, careful adjustment of anesthesia is needed.
3. To optimize oral gavage, hold the neck of a mouse straight so that a feeding needle is easily inserted into the esophagus through the pharynx.
4. While 5 h is recommended between gavage and blood collection for sepsis models, researchers have used 3 h in a murine colitis model [9]. The rationale for increasing the time in sepsis relates to decreased gut motility in critical illness.
5. Creatinine concentration after gavage is significantly higher than endogenous creatinine concentration; therefore, endogenous creatinine has a negligible effect on the results obtained.
6. It is possible to measure FD4 and creatinine in an everted gut sac model at the same time.
7. The segment of the intestine examined should be determined depending on the research interest or the purpose of the project. We typically prepare 10 cm of jejunum, ileum, and colon when measuring permeability [11].
8. A hole in an everted gut sac reduces the accuracy of permeability measurement. When removing the intestine from surrounding fat and mesentery, meticulous attention must be paid to surgical technique so as not to injure the intestine.
9. If the everted gut sac leaks after injection of the fluid into the ligated gut lumen, the experiment should be terminated and a different gut sac should be examined in order to obtain consistent results.

10. Although the length of the intestine is accounted for in the formula to calculate the final result, it is more difficult to inflate the gut equally with KHBB when longer segments of intestine are obtained, affecting the accuracy of the measurement. Alternatively, when examining a short intestinal segment, you can use the actual volume injected to calculate permeability if unable to inject the entire volume that was planned.
11. To separate mucus and epithelial cells, the sample collected from an everted gut sac should be centrifuged prior to measuring fluorescent activity.

Acknowledgments

The authors declare no conflict of interests related to the current study. This work was supported by the National Institutes of Health under grants (GM072808, AA027396, GM104323, GM113228).

References

1. Meng M, Klingensmith NJ, Coopersmith CM (2017) New insights into the gut as the driver of critical illness and organ failure. Curr Opin Crit Care 23:143–148
2. Mittal R, Coopersmith CM (2014) Redefining the gut as the motor of critical illness. Trends Mol Med 20:214–223
3. Lorentz CA, Liang Z, Meng M et al (2017) Myosin light chain kinase knockout improves gut barrier function and confers a survival advantage in polymicrobial sepsis. Mol Med 23:155–165
4. Clark JA, Clark AT, Hotchkiss RS et al (2008) Epidermal growth factor treatment decreases mortality and is associated with improved gut integrity in sepsis. Shock 30:36–42
5. Turner JR (2009) Intestinal mucosal barrier function in health and disease. Nat Rev Immunol 9:799–809
6. Garcia-Hernandez V, Quiros M, Nusrat A (2017) Intestinal epithelial claudins: expression and regulation in homeostasis and inflammation. Ann N Y Acad Sci 1397:66–79
7. Tsukita S, Tanaka H, Tamura A (2019) The Claudins: from tight junctions to biological systems. Trends Biochem Sci 44:141–152
8. Odenwald MA, Turner JR (2017) The intestinal epithelial barrier: a therapeutic target? Nat Rev Gastroenterol Hepatol 14:9–21
9. Tsai PY, Zhang B, He WQ et al (2017) IL-22 upregulates epithelial Claudin-2 to drive diarrhea and enteric pathogen clearance. Cell Host Microbe 21:671–681. e4
10. Alam MA, Al-Jenoobi FI, Al-Mohizea AM (2012) Everted gut sac model as a tool in pharmaceutical research: limitations and applications. J Pharm Pharmacol 64:326–336
11. Yoseph BP, Klingensmith NJ, Liang Z et al (2016) Mechanisms of intestinal barrier dysfunction in sepsis. Shock 46:52–59

Chapter 16

Sepsis Biomarkers

Yachana Kataria and Daniel Remick

Abstract

Biomarkers have been used in sepsis to assist with the diagnosis of disease as well as determining the severity of disease, that is, prognosis. These biomarkers are based on the presence of discrete molecules within the blood. Unfortunately, in 2020, a single biomarker does not have sufficient sensitivity and specificity to definitively rule in or rule out sepsis. Biomarkers have shown better performance in animal models of disease.

Key words Sepsis, C-reactive protein, IL-6, Procalcitonin, Antibiotics

1 Introduction

1.1 Sepsis

Sepsis kills and we do not know why. Basic scientists do not understand the immunopathology of the disease and clinicians struggle to treat patients. The impact of sepsis is staggering with an estimated global burden of 31.5 million cases and 5.3 million deaths each year. While the surviving sepsis campaign instituted bundles of care protocols, the decrease in mortality was only modest [1, 2]. A meta-analysis of two decades of sepsis therapies only showed a slow decrease in mortality over time which is not statistically significant [3]. The problem continues beyond the first month since 40% of septic patients will die within the first year [4, 5].

Sepsis it is a heterogeneous clinical condition from the perspective of immunologic and physiologic responses. A concept emerged in 1992 that the septic patients have a systemic inflammatory response syndrome (SIRS) [6]. A patient was considered septic if there was a suspected infection and two of the four SIRS criteria were present. The nonspecific sepsis criteria were felt to be necessary to allow the rapid enrollment of septic patients in clinical trials, before laboratory tests would be available. However, the lack of specificity created tremendous heterogeneity among the patients in multiple parameters, such as the concentrations of pro-inflammatory cytokines, changes in types of white blood cells,

Wendy E. Walker (ed.), *Sepsis: Methods and Protocols*, Methods in Molecular Biology, vol. 2321,
https://doi.org/10.1007/978-1-0716-1488-4_16,

physiologic responses and the presence of organ injury. Each of these parameters may change over time [7]. The 2016 definition of sepsis states that sepsis is life threatening organ dysfunction caused by a dysregulated host response to infection [8].

Given the enormous problem of sepsis, patient care could be improved if there were a laboratory test, or biomarker, that would accurately diagnose sepsis. In the current chapter we review the proposed biomarkers for sepsis.

2 Biomarkers in Sepsis

2.1 Biomarkers Overview

The National Institutes of Health Biomarkers Definitions Working Group defined a biomarker as "a characteristic that is objectively measured and evaluated as an indicator of normal biological processes, pathogenic processes or pharmacologic responses to a therapeutic intervention." In the context of this definition, biomarkers may be considered as potential indicators for susceptibility to disease, to measure biological responses, to identify early disease states and to characterize different types of disease states. Biomarkers are tools to help diagnose, provide risk stratification, or monitor disease. Biomarkers may not actually cause the disease and in several instances will serve as a surrogate for the disease. A single biomarker can have more than one application. Over the last three decades there have been many proposed sepsis biomarkers for applications in diagnosis, prognosis and antibiotic guidance. While many biomarkers have been proposed, interleukin 6 (IL-6), C-reactive protein (CRP) and procalcitonin (PCT) are the most widely studied. As such, we examine and assess the most recent studies that evaluate their clinical utility in diagnosis, prognosis and guidance for antibiotic therapy in sepsis. Extensive reviews of these biomarkers and others can be found elsewhere [9–11].

2.2 IL-6—Basic Biology

IL-6 is a proinflammatory and anti-inflammatory cytokine regulating immune responses, hematopoiesis and playing a role in inflammation. IL-6 is composed of 184 amino acids with a molecular weight of 21–26 kDa, and its gene is located on chromosome 7 [12, 13]. The IL-6 receptor is found on several cell types, such as B and T lymphocytes, and monocytes/macrophages. It stimulates T and B lymphocytes and other cells involved in cell proliferation and differentiation and also enhances their functionality [14]. Typically, normal serum concentrations have been found to be less than 5 pg/mL. In contrast, in patients with sepsis, IL-6 levels are reported to be greater than 500 pg/mL [15]. IL-6 also induces synthesis of acute phase proteins in the liver, such as C reactive protein.

2.3 C-Reactive Protein—Basic Biology

C-reactive protein (CRP) is synthesized by hepatocytes in response to stimulation by cytokines (i.e., IL-6). It is an acute phase reactant protein widely used to monitor noninfectious inflammatory diseases. As a diagnostic marker, CRP rises slowly at the onset of sepsis in contrast to IL-6 which peaks around 36 h after an endotoxin challenge [16, 17]. CRP exhibits a half-life of approximately 19 h. Increases in serum CRP concentrations in septic patients are associated with increased risk of organ failure and death (36 vs. 21%, $p < 0.05$) [18]. CRP at normal physiological levels is found to be less than 5 μg/mL. However, its levels reach as high as 500 μg/mL during inflammation [19].

2.4 Procalcitonin—Basic Biology

Procalcitonin (PCT) is a 116-peptide precursor hormone of calcitonin involved in calcium homeostasis. PCT is synthesized by the C-cells in the thyroid. However, during sepsis PCT synthesis is produced by extrathyroidal neuroendocrine tissue. Like CRP, PCT is also considered an acute phase protein. PCT levels are elevated as part of the inflammatory response to trauma, major surgery, and pancreatitis. In contrast to CRP, elevations of PCT also occur in renal impairment making it a less specific biomarker in renal disease [20]. Typically, PCT levels are less than 0.1 ng/mL. PCT has a longer window of detection as it exhibits a half-life of 24 h.

2.5 Biomarkers for Diagnosis of Sepsis

The gold standard criterion for diagnosis of infection is microbiological cultures. Only 30–40% of septic patients have a positive culture as microorganisms are not always detectable in the blood [8]. The turnaround time for microbiological culture is typically over a day due to the necessary incubation time. This highlights the importance of having a biomarker with improved accuracy that has a quick turnaround time in a cost-effective manner. The clinical presentation of sepsis is also not specific (i.e., tachycardia, raised white cell count, fever, etc.) since these signs and symptoms can be commonly found in other diseases as well. Therefore, when assessing the diagnostic performance of IL-6, CRP, and PCT, it is important to assess the specificity of the test which would help determine the presence of sepsis.

IL-6 is present in the blood earlier than CRP or PCT [21]. Serum levels of IL-6 have been shown to rise as early as 1 h; they peak within 2 h after the onset of stimulus of the infection [22]. In contrast, serum PCT levels start to rise typically in 3–6 h, and peak between 8 and 24 h, in response to a bacterial infection. CRP also rises slowly at the onset of sepsis, typically after 12–24 h, and peaks around 36 h [17]. The earlier rise in PCT makes it a good candidate for a diagnostic biomarker. Additionally, it has been demonstrated that PCT levels do not stay elevated as long as CRP.

IL-6 has been shown to be upregulated in septic patients and exhibits higher sensitivity and specificity as a diagnostic biomarker relative to other cytokines [23]. IL-6 levels are also correlated with mortality where increased IL-6 levels were found to be associated with the higher risk of death in patients with sepsis [24]. Of note, IL-6 levels decrease after the exposure to bacterial products in the early stages of infection [25]. A few meta-analyses have assessed the diagnostic accuracy of IL-6 [26–28]. A meta-analysis aimed to evaluate the role of IL-6 in differentiating sepsis from noninfectious systemic inflammatory response syndrome (SIRS) concluded that IL-6 had moderate diagnostic performance based on a summary area under the receiver operating characteristic (ROC) curve of 0.80 [28]. This paper also reported a pooled sensitivity of 0.68 (95% CI: 0.65–0.70) and a specificity of 0.73 (95% CI: 0.71–0.76). A more recent meta-analysis observed a similar sensitivity and specificity of IL-6 in critically ill patients [29].

CRP, relative to the IL-6 and PCT, remains the most widely available and accessible test. Diagnostic assessment of CRP in the ICU patient population has yielded sensitivity and specificity of 71.8% and 66.6% at the cutoff of 7.9 mg/L. However, CRP is not a specific marker for sepsis making it less than an ideal biomarker. A meta-analysis evaluated the role of IL-6, PCT, and CRP in the context of differentiating sepsis from noninfections systemic inflammatory response syndrome and found that the area under the curve was found to be 0.80, 0.83 and 0.71, respectively suggesting that IL-6 and PCT have better diagnostic value compared to CRP [26].

When comparing the diagnostic accuracy of PCT and CRP there is a trade-off between sensitivity and specificity. A recent study evaluating the diagnostic accuracy of PCT and CRP in patients with suspected bacterial sepsis side by side utilizing the Sepsis-3 criteria found that CRP (cutoff: 20 mg/L) had a higher sensitivity at 88.4% (95% CI: 85.8–91.1) compared to PCT (cutoff 2.0 ng/mL) at 32.1% (95% CI: 28.3–36.0) [30]. In contrast, PCT exhibited a higher specificity (97.4% (95% CI: 96.5–98.4) vs. 14.4% (95% CI:12.2–16.5)) [30]. They also assessed the diagnostic performance of a composite score utilizing PCT, CRP, and lactate levels from a linear discriminant analysis. The composite score was able to achieve an AUC of 0.85 (95% CI: 0.83–0.89) [30]. While a composite score might appeal at first glance, it has its own limitations in regards to implementation in the clinic, emergency department or intensive care unit.

Under different clinical contexts, PCT and CRP exhibit different diagnostic performance characteristics. CRP is reportedly a better biomarker in patients with acute exacerbation of chronic obstructive pulmonary disease or in patients with suspected acquired pneumonia [31, 32]. There is evidence that PCT can differentiate bacterial pneumonia from noninfectious pulmonary pathologies [33, 34]. Procalcitonin levels are also increased in non-infectious inflammatory conditions such as patients with thyroid carcinoma, after cardiac surgery, and following cardiac arrest [20, 35]. Since PCT levels are also elevated in patients with chronic kidney disease, a different diagnostic cutoff value needs to be utilized in this patient population [36].

A systematic review of 59 studies evaluating the ability of sepsis biomarkers to distinguish the types of infection, specifically, bacterial from nonbacterial infections, found that the studies were less than ideal [37]. Of all the biomarkers examined in this study, CRP and PCT tended to be higher in patients with bacterial growth than those with nonbacterial infections. They found that the sensitivity and specificity for CRP ranged from 61.2% to 100% and from 26% to 100%, respectively. The sensitivity and specificity for PCT ranged from 38% to 97% and from 31% to 100%, respectively, to identify bacterial infections [37]. Other studies suggested that low procalcitonin levels can be utilized to rule out bacteremia, particularly in ICU patients [34, 38].

In neonates, procalcitonin has shown a high negative predictive value for invasive bacterial infections [39, 40]. However, PCT lacks specificity in the first 48 h of life due to physiological increases. The interpretation of PCT values is further complicated by perinatal factors such as maternal preeclampsia and respiratory distress syndrome [41]. Some small studies have reported that the CRP diagnostic performance is comparable to PCT in neonatal sepsis [42]. Recently, a randomized controlled clinical trial reported that PCT guided decisions for the length of antibiotic usage led to a low rate of infection relative to standard of care [43].

2.6 Biomarkers for Prognosis

IL-6, CRP, and PCT have also been reported to have prognostic capabilities; more specifically the biomarkers may elucidate the severity of disease which can help with risk stratification of patients. It was initially reported that IL-6 levels in septic patients are correlated with mortality [44, 45] and other studies found that the mortality of patients admitted to the ICU correlated with IL-6 levels [45, 46]. In another study, patients were stratified by IL-6 levels to receive immunomodulating therapy but the mortality rates between the groups were not found to be statistically significant [16]. Blood IL-6 levels were also found to be positively correlated with the quick Sequential Organ Failure Assessment (qSOFA) score in patients with SIRS [47, 48]. These findings indicate that IL-6 may have a role as a prognostic biomarker in septic patients.

Elevated CRP levels are also associated with poor long-term outcomes and mortality for septic patients. For example, ICU patients with CRP levels greater than or equal to 10 mg/dL are more likely develop organ dysfunction in comparison to ICU patients who had CRP concentrations less than 1 mg/dL

[18]. The study also reported that a decrease in CRP level after 48 h was associated with a mortality rate of 15.4%, whereas CRP levels that remained greater than 10 mg/dL had a mortality rate of 60.9% [18]. Another study assessing the prognostic utility of CRP in patients with clinically suspected sepsis found that serum CRP levels were an independent predictor of death with an adjusted odds ratio of 2.0 (95% CI: 1.3–2.1) [19].

Several studies, but not all, have demonstrated that serum PCT levels are correlated with the severity of sepsis and organ dysfunction. In a cohort of 234 ICU patients with sepsis, those with a PCT level greater than >0.85 ng/mL had significantly higher mortality rate of 45% vs. 26% [49]. One study found that serum PCT levels upon admission to the ED report an area under the curve (AUC) of 0.686 (95% CI: 0.613–0.752) for death within 30 days, with a sensitivity and specificity of 0.96 (95% CI: 0.90–0.98) and 0.73 (95% CI: 0.52–0.88), respectively [50]. A recent meta-analysis found that elevated PCT concentrations were associated with higher risk of death in all patients; they reported a pooled relative risk of 2.60 (95% CI; 2.05–3.30) from a random effects model [51]. They also evaluated nonclearance of PCT in predicting sepsis mortality since patients at a high risk of dying may have persistently elevated PCT levels. They reported a pooled relative risk of 3.05 (95% CI: 2.35–3.95). The overall area under the curve was 0.79 (95% CI: 0.75–0.83) with a sensitivity and specificity of 0.72% (95% CI: 0.58–0.82) and 0.77% (95% CI: 0.55–0.90), respectively. Collectively suggesting that nonclearance PCT is associated with mortality in septic patients. Karlsson has reported that changes in PCT concentrations in the first 72 hours and not initial concentrations were associated with mortality [52]. More specifically, mortality rates were significantly lower in patients whose PCT levels decrease less than 50% over 72 h in comparison to individuals whose levels decreased more than 50%. These findings are corroborated with other smaller studies [53–55]. Most recently, a multicenter prospective study across the United States found that failure to decrease PCT levels by 80% by the fourth day was an independent predictor of 28-day mortality (Hazard Ratio 1.95; 95% CI: 1.18–3.30) [56].

Only a few studies have assessed CRP and PCT side by side and their findings should be interpreted cautiously as they are relatively small studies. Ryu et al. assessed the clinical utility of procalcitonin and CRP as outcome predictors in critically ill patients with severe sepsis and reported that PCT and CRP predicted 28-day mortality rates in a comparable fashion [57]. Moreover, they claim that CRP may be equivalent to PCT in predicting outcomes in these patients. In contrast, Hoeboer et al. reported the PCT was better than CRP to predict mortality and septic shock in critically patients [58].

2.7 Antibiotic Guidance

Changes in biomarker levels throughout the course of the disease could elucidate whether patients might benefit from appropriate antibiotic therapy, that is, early termination. These biomarkers have been shown to guide antimicrobial therapy, the so-called antibiotic stewardship, since it is important to prevent unnecessary antibiotic prescriptions while ensuring efficacy. Antimicrobial resistance is becoming a growing public health concern due to the increase use of broad-spectrum antibiotics. As such, early and appropriate antibiotic therapy has been shown to improve outcomes in patients with sepsis.

PCT continues to be widely studied for antibiotic stewardship and it has been demonstrated that serum PCT levels decrease with the resolution of sepsis and antibiotic therapy [52, 59, 60]. Antibiotic therapy based on PCT levels has been studied in patients with lower respiratory tract infections, critically ill sepsis patients, as well as patients with urinary tract infections, postoperative infections, meningitis, and acute heart failure with potential superinfection (i.e., pneumonia) [61]. There is strong evidence for PCT guided antibiotic stewardship in respiratory infection and sepsis, but it is not as well studied in other diseases.

A large RCT from the Netherlands assessed the utility of PCT guided antibiotic treatment in the ICU reported that PCT decreases the duration of antimicrobial therapy by 2 days in critically ill patients with a presumed bacterial infection [34]. Furthermore, they also reported that the decrease in days of antibiotic treatment was also associated with a decrease in 28-day mortality (25% vs. 20%; $p = 0.0122$). Another RCT assessing the effect of PCT in patients with severe sepsis or septic shock found that while PCT reduced antimicrobial exposure, it did not significantly reduce mortality rates [62].

Schuetz et al. conducted a recent Cochrane review which included 26 randomized control trials investigating the effect of procalcitonin guided antibiotic treatment on mortality in acute respiratory tract infections [63–65]. They found that PCT guidance reduced antibiotic exposure by 2.4 days (5.7 days vs. 8.1 days; 95% CI: −2.71 to −2.15). In addition, they reported a significant difference in mortality rates (adjusted odds ratio 0.83; 95% CI: 0.70–0.99). Another meta-analysis evaluated the PCT guided treatment in acute exacerbation of chronic obstructive pulmonary disease involving 4 trials and 679 patients. This study found that PCT level-guided treatment significantly reduced antibiotic use relative to the standard treatment, without increasing mortality [66]. In contrast, another recent meta-analysis by Andriolo et al. studied the effectiveness and safety of PCT evaluation for reducing morality in adults with sepsis, severe sepsis, or septic shock across 10 RCT trials with 1215 participants and did not find clear evidence to "support the use of procalcitonin-guided antimicrobial therapy to minimize mortality, mechanical ventilation, clinical severity, reinfection or duration of antimicrobial therapy of patients with septic conditions." [67]. While there is no consensus regarding mortality rates there is strong evidence that PCT does decrease the days of antimicrobial therapy. These incongruent results might be attributable to the lack of consensus regarding utilizing PCT for antibiotic stewardship, and no consensus in regards to the best cutoff level for PCT.

There have been no studies to date assessing the role of IL-6 in antibiotic stewardship. However, there have been limited studies evaluating the role of CRP in therapeutic guidance. Schmit et al. found that changes of CRP levels over the initial 2 days of therapy can help assess the response to antimicrobial therapy in septic patients [68–70]. Likewise, CRP levels have been found to be lower in septic patients who had a more favorable response to pneumonia treatment and vice versa [69, 71]. Most recently, a meta-analysis published that the use of CRP to guide the use of antibiotics was associated with reduced antibiotic initiation in adult outpatients and in neonatal patients [72]. The recommended cutoff thresholds were different for both of the groups. For the adult outpatients, the recommended CRP cutoff for withholding or treatment initiation are <20 mg/L and ≥ 100 mg/L, respectively. In contrast, they recommended a CRP level of <10 mg/L for discontinuing antibiotic treatment in neonates. Olivera et al. studied the antibiotic stewardship role of PCT and CRP in ICU patients with sepsis and found that they both have similar characteristics in guiding antibiotic therapy [70].

2.8 Biomarkers in Animal Models of Sepsis

Biomarkers have also been used to predict mortality in animal models of sepsis. The most widely used animal model of sepsis is cecal ligation and puncture, either in mice or in rats [73]. A 2002 paper detailed how plasma levels of IL-6 obtained within the first 6 h of sepsis predicted subsequent early mortality [74]. The findings were confirmed in later publications [75–77]. The plasma level of IL-6 could be used to stratify mice into high or low mortality groups. With this stratification, interventions are only effective in the high mortality group [78].

The plasma levels of IL-6 correlate with the murine septic phenotype, with higher levels found in mice with higher mortality [79]. It should be noted that mice with extremely high IL-6 levels (>14,000 pg/mL) administering antibiotics will not decrease mortality, although the antibiotics were effective in mice with lower levels of IL-6 [80].

3 Concluding Remarks

Sepsis is a complex clinical condition and presents as a heterogeneous phenotype in affected patients. An ideal biomarker for sepsis would have high sensitivity and specificity and would be available with fully automated technology with a short turnaround time. The biomarker must also be also be easily interpretable and minimally invasive.

Elevated levels of troponin are widely used to diagnose heart attacks, but there has been no equivalent of troponin identified for sepsis. Biomarkers for sepsis are far from perfect and much work remains to accurately elucidate the biologically heterogeneous process. Sepsis biomarkers have been primarily used for three different applications, diagnosis, prognosis, and guidance of antibiotic therapy. IL-6, CRP, and PCT cannot be utilized solely for diagnosis, prognosis, or antibiotic stewardship; rather, it must be used in conjunction with the entire clinical picture. None of them have sufficient sensitivity and specificity. PCT appears to be more specific and a better prognostic biomarker than CRP and IL-6. In regards to antibiotic stewardship, there is moderate evidence suggesting the use of PCT. Again, this emphasizes the clinical status and evolution of the patient when they present with symptom related to sepsis.

Collectively there have been inconsistent and heterogeneous results regarding the performance of these biomarkers; the findings invariably depend on severity and type of infection. Most studies conducted to date have limitations due to small sample sizes, different clinical settings, different assays, and different cutoff thresholds, making comparison of results difficult. PCT is not widely available as a test and is costly compared to CRP. These various factors contribute to the observed heterogeneity in the studies. It is plausible that combining these biomarkers might improve their performance although this concept has not been extensively studied. A few selected studies have examined combinations of biomarkers and these have been promising but need further investigation [30, 81]. It is also plausible that the findings to date reflect the complexity of sepsis.

Regardless of the limitations, there is still clinical utility in these biomarkers. The presentation and site of infection will vary among people and as such biomarker levels to guide therapy will vary. The underlying pathophysiology of sepsis is complex, emphasizing the importance of the whole clinical picture. While it remains unlikely to find a silver bullet for a highly complex disease there are new promising candidate biomarkers that are currently being evaluated.

Acknowledgments

Supported in part by NIH grants T32GM86308, R01GM117519, and R01HL GM 86308-06.

References

1. Levy MM, Rhodes A, Phillips GS et al (2014) Surviving Sepsis Campaign: association between performance metrics and outcomes in a 7.5-year study. Crit Care Med 40:1623–1633
2. Fleischmann C, Scherag A, Adhikari NK et al (2016) Assessment of global incidence and mortality of hospital-treated sepsis. Current estimates and limitations. Am J Respir Crit Care Med 193:259–272
3. Stevenson EK, Rubenstein AR, Radin GT et al (2014) Two decades of mortality trends among patients with severe sepsis: a comparative meta-analysis. Crit Care Med 42:625
4. Rhee C, Dantes R, Epstein L et al (2017) Incidence and trends of sepsis in US hospitals using clinical vs. claims data, 2009-2014. JAMA 318:1241–1249
5. Medicine PIJNEJo \ (2014) A randomized trial of protocol-based care for early septic shock N Engl J Med 370:1683–1693
6. Physicians ACoC,Med SoCCMCCCJCC (1992) American College of Chest Physicians/Society of Critical Care Medicine Consensus Conference: definitions for sepsis and organ failure and guidelines for the use of innovative therapies in sepsis. Crit Care Med 20:864–874
7. Stortz JA, Murphy TJ, Raymond SL et al (2018) Evidence for persistent immune suppression in patients who develop chronic critical illness after sepsis. Shock 49:249–258
8. Singer M, Deutschman CS, Seymour CW et al (2016) The third international consensus definitions for sepsis and septic shock (Sepsis-3). JAMA 315:801–810
9. van Engelen TS, Wiersinga WJ, Scicluna BP et al (2018) Biomarkers in sepsis. Critical Care Clinics 34:139–152
10. Prucha M, Bellingan G, Zazula R (2015) Sepsis biomarkers. Clin Chim Acta 440:97–103
11. Pierrakos C, Vincent J-L (2010) Sepsis biomarkers: a review. Crit Care 14:R15
12. Sibbing D, Koch W, Massberg S et al (2011) No association of paraoxonase-1 Q192R genotypes with platelet response to clopidogrel and risk of stent thrombosis after coronary stenting. Eur Heart J 32:1605–1613
13. Hirano T (1991) Interleukin 6 (IL-6) and its receptor: their role in plasma cell neoplasias. Int J Cell Cloning 9:166–184
14. Sims JE, Smith DE (2010) The IL-1 family: regulators of immunity. Nat Rev Immunol 10:89
15. Bloos F, Reinhart K (2014) Rapid diagnosis of sepsis. Virulence 5:154–160
16. Reinhart K, Menges T, Gardlund B et al (2001) Randomized, placebo-controlled trial of the anti-tumor necrosis factor antibody fragment afelimomab in hyperinflammatory response during severe sepsis: The RAMSES Study. Crit Care Med 29:765–769
17. Shehabi Y, Seppelt IJCC (2008) Pro/Con debate: is procalcitonin useful for guiding antibiotic decision making in critically ill patients? Crit Care 12:211
18. Lobo SM, Lobo FR, Bota DP et al (2003) C-reactive protein levels correlate with mortality and organ failure in critically Ill patientsa. Chest 123:2043–2049
19. Yamamoto S, Yamazaki S, Shimizu T et al (2015) Prognostic utility of serum CRP levels in combination with CURB-65 in patients with clinically suspected sepsis: a decision curve analysis. PLoS One 5:e007049
20. Reinhart K, Karzai W, Meisner MJ (2000) Procalcitonin as a marker of the systemic inflammatory response to infection. Intensive Care Med 26:1193–1200
21. Dahaba A, Metzler HJ (2009) Procalcitonin's role in the sepsis cascade. Is procalcitonin a sepsis marker or mediator? Minerva Anestesiol 75:447–452
22. Song M, Kellum JA (2005) Interleukin-6. Crit Care Med 33:S463–S465
23. Feng M, Sun T, Zhao Y et al (2016) Detection of serum Interleukin-6/10/18 levels in sepsis and its clinical significance. J Clin Lab Anal 30:1037–1043
24. Mera S, Tatulescu D, Cismaru C et al (2011) Multiplex cytokine profiling in patients with sepsis. APMIS 119:155–163

25. Jekarl DW, Lee S-Y, Lee J et al (2013) Procalcitonin as a diagnostic marker and IL-6 as a prognostic marker for sepsis. Diagn Microbiol Infect Dis 75:342–347
26. Ma L, Zhang H, Y-l Y et al (2016) Role of interleukin-6 to differentiate sepsis from non-infectious systemic inflammatory response syndrome. Cytokine 88:126–135
27. Liu Y, Hou J-H, Li Q et al (2016) Biomarkers for diagnosis of sepsis in patients with systemic inflammatory response syndrome: a systematic review and meta-analysis. Springerplus 5:2091
28. Hou T, Huang D, Zeng R et al (2015) Accuracy of serum interleukin (IL)-6 in sepsis diagnosis: a systematic review and meta-analysis. Int J Clin Exp Med 8:15238
29. Iwase S, Nakada T-A, Hattori N et al (2018) Interleukin-6 as a diagnostic marker for infection in critically ill patients: A systematic review and meta-analysis. Am J Emerg Med 37 (2):260–265
30. Ljungström L, Pernestig A-K, Jacobsson G et al (2017) Diagnostic accuracy of procalcitonin, neutrophil-lymphocyte count ratio, C-reactive protein, and lactate in patients with suspected bacterial sepsis. PLoS One 12: e0181704
31. Daniels JM, Schoorl M, Snijders D et al (2010) Procalcitonin vs C-reactive protein as predictive markers of response to antibiotic therapy in acute exacerbations of COPD. Chest 138:1108–1115
32. Gaïni S, Koldkjær OG, Pedersen C, Pedersen SS (2006) Procalcitonin, lipopolysaccharide-binding protein, interleukin-6 and C-reactive protein in community-acquired infections and sepsis: a prospective study. Crit Care 10:R53
33. Schuetz P, Kutz A, Grolimund E et al (2014) Excluding infection through procalcitonin testing improves outcomes of congestive heart failure patients presenting with acute respiratory symptoms: results from the randomized ProHOSP trial. Int J Cardiol 175:464–472
34. de Jong E, van Oers JA, Beishuizen A et al (2016) Efficacy and safety of procalcitonin guidance in reducing the duration of antibiotic treatment in critically ill patients: a randomised, controlled, open-label trial. Lancet Infect Dis 16:819–827
35. Schuetz P, Affolter B, Hunziker S et al (2010) Serum procalcitonin C-reactive protein and white blood cell levels following hypothermia after cardiac arrest: a retrospective cohort study. Eur J Clin Invest 40:376–381
36. Grace E, Turner RM (2014) Use of procalcitonin in patients with various degrees of chronic kidney disease including renal replacement therapy. Clin Infect Dis 59:1761–1767
37. Kapasi AJ, Dittrich S, González IJ et al (2016) Host biomarkers for distinguishing bacterial from non-bacterial causes of acute febrile illness: a comprehensive review. PLoS One 11: e0160278
38. Hoeboer SH, van der Geest PJ, Nieboer D et al (2015) The diagnostic accuracy of procalcitonin for bacteraemia: a systematic review and meta-analysis. Clin Microbiol Infect 21:474–481
39. Van Rossum A, Wulkan RW, Oudesluys-Murphy AM (2004) Procalcitonin as an early marker of infection in neonates and children. Lancet Infect Dis 4:620–630
40. Vouloumanou EK, Plessa E, Karageorgopoulos DE et al (2011) Serum procalcitonin as a diagnostic marker for neonatal sepsis: a systematic review and meta-analysis. Intensive Care Med 37:747–762
41. Chiesa C, Pellegrini G, Panero A et al (2003) C-reactive protein, interleukin-6, and procalcitonin in the immediate postnatal period: influence of illness severity, risk status, antenatal and perinatal complications, and infection. Clin Chem 49:60–68
42. Boraey N, Sheneef A, Mohammad MAA et al (2012) Procalcitonin and C-reactive protein as diagnostic markers of neonatal sepsis. Ginekol Pol 6:e14
43. Stocker M, van Herk W, el Helou S et al (2017) Procalcitonin-guided decision making for duration of antibiotic therapy in neonates with suspected early-onset sepsis: a multicentre, randomised controlled trial (NeoPIns). Lancet 390:871–881
44. Hack CE, De Groot ER, Felt-Bersma R et al (1989) Increased plasma levels of interleukin-6 in sepsis. Blood 74:1704–1710
45. Waage A, Brandtzaeg P, Halstensen A et al (1989) The complex pattern of cytokines in serum from patients with meningococcal septic shock. Association between interleukin 6, interleukin 1, and fatal outcome. J Exp Med 169:333–338
46. Friedland JS, Porter JC, Daryanani S et al (1996) Plasma proinflammatory cytokine concentrations, Acute Physiology and Chronic Health Evaluation (APACHE) III scores and survival in patients in an intensive care unit. Crit Care Med 24:1775–1781
47. Oda S, Hirasawa H, Shiga H et al (2005) Sequential measurement of IL-6 blood levels in patients with systemic inflammatory

response syndrome (SIRS)/sepsis. Cytokine 29:169–175

48. Mat-Nor MB, Ralib AM, Abdulah NZ et al (2016) The diagnostic ability of procalcitonin and interleukin-6 to differentiate infectious from noninfectious systemic inflammatory response syndrome and to predict mortality. J Crit Care 33:245–251
49. Giamarellos-Bourboulis E, Tsangaris I, Kanni T et al (2011) Procalcitonin as an early indicator of outcome in sepsis: a prospective observational study. J Hosp Infect 77:58–63
50. Peschanski N, Chenevier-Gobeaux C, Mzabi L et al (2016) Prognostic value of PCT in septic emergency patients. Ann Intensive Care 6:47
51. Liu D, Su L, Han G et al (2015) Prognostic value of procalcitonin in adult patients with sepsis: a systematic review and meta-analysis. PLoS One 10:e0129450
52. Karlsson S, Heikkinen M, Pettilä V et al (2010) Predictive value of procalcitonin decrease in patients with severe sepsis: a prospective observational study. Crit Care 14:R205
53. Kutz A, Mueller B, Schuetz PJCC (2015) Prognostic value of procalcitonin in respiratory tract infections across clinical settings. Crit Care 19:P65
54. Ruiz-Rodríguez J, Caballero J, Ruiz-Sanmartin A et al (2012) Usefulness of procalcitonin clearance as a prognostic biomarker in septic shock. A prospective pilot study. Med Intensiva 36:475–480
55. Pieralli F, Vannucchi V, Mancini A et al (2015) Procalcitonin kinetics in the first 72 hours predicts 30-day mortality in severely ill septic patients admitted to an intermediate care unit. J Clin Med Res 7:706
56. Schuetz P, Birkhahn R, Sherwin R et al (2017) Serial procalcitonin predicts mortality in severe sepsis patients: results from the multicenter procalcitonin monitoring sepsis (MOSES) study. Crit Care Med 45:781
57. Ryu J-A, Yang JH, Lee D et al (2015) Clinical usefulness of procalcitonin and C-reactive protein as outcome predictors in critically ill patients with severe sepsis and septic Shock. PLoS One 10:e0138150
58. Hoeboer S, Groeneveld ABJ (2013) Changes in circulating procalcitonin versus C-reactive protein in predicting evolution of infectious disease in febrile, critically ill patients. PLoS One 17:30
59. Schuetz P, Maurer P, Punjabi V et al (2013) Procalcitonin decrease over 72 hours in US critical care units predicts fatal outcome in sepsis patients. Crit Care 17:R115
60. Christ-Crain M, Jaccard-Stolz D, Bingisser R et al (2004) Effect of procalcitonin-guided treatment on antibiotic use and outcome in lower respiratory tract infections: cluster-randomised, single-blinded intervention trial. Lancet 363:600–607
61. Sager R, Kutz A, Mueller B et al (2017) Procalcitonin-guided diagnosis and antibiotic stewardship revisited. BMC Med 15:15
62. Bloos F, Trips E, Nierhaus A et al (2016) Effect of sodium selenite administration and procalcitonin-guided therapy on mortality in patients with severe sepsis or septic shock: a randomized clinical trial. JAMA Intern Med 176:1266–1276
63. Schuetz P, Muller B, Christ-Crain M et al (2013) Procalcitonin to initiate or discontinue antibiotics in acute respiratory tract infections. Cochrane Database Syst Rev 8:1297–1371
64. Schuetz P, Wirz Y, Sager R et al (2018) Effect of procalcitonin-guided antibiotic treatment on mortality in acute respiratory infections: a patient level meta-analysis. Lancet Infect Dis 18:95–107
65. Schuetz P, Bretscher C, Bernasconi L et al (2017) Overview of procalcitonin assays and procalcitonin-guided protocols for the management of patients with infections and sepsis. Expert Rev Mol Diagn 17:593–601
66. Lin C, Pang Q (2018) Meta-analysis and systematic review of procalcitonin-guided treatment in acute exacerbation of chronic obstructive pulmonary disease. Clin Respir J 12:10–15
67. Andriolo BN, \Andriolo RB, Salomao R et al (2017) Effectiveness and safety of procalcitonin evaluation for reducing mortality in adults with sepsis, severe sepsis or septic shock Cochrane Database Syst Rev 1(1):CD010959
68. Schmit X, Vincent JLJI (2008) The time course of blood C-reactive protein concentrations in relation to the response to initial antimicrobial therapy in patients with sepsis. Infection 36:213–219
69. Lisboa T, Seligman R, Diaz E et al (2008) C-reactive protein correlates with bacterial load and appropriate antibiotic therapy in suspected ventilator-associated pneumonia. Crit Care Med 36:166–171
70. Oliveira CF, Botoni FA, Oliveira CR et al (2013) Procalcitonin versus C-reactive protein for guiding antibiotic therapy in sepsis: a randomized trial. Crit Care Med 41:2336–2343
71. Bruns AH, Oosterheert JJ, Hak E et al (2008) Usefulness of consecutive CRP measurements in follow-up of treatment for severe CAP. Crit Care 11:R92

72. Petel D, Winters N, Gore GC et al (2018) Use of C-reactive protein to tailor antibiotic use: a systematic review and meta-analysis. BMJ Open 8:e022133
73. Osuchowski MF, Ayala A, Bahrami S et al (2018) Minimum quality threshold in pre-clinical sepsis studies (MQTiPSS): an international expert consensus initiative for improvement of animal modeling in sepsis. Shock 50:377–380
74. Remick DG, Bolgos GR, Siddiqui J et al (2002) Six at six: interleukin-6 measured 6 h after the initiation of sepsis predicts mortality over 3 days. Shock 17:463–467
75. Craciun FL, Iskander KN, Chiswick EL et al (2014) Early murine polymicrobial sepsis predominantly causes renal injury. Shock 41:97–103
76. Craciun FL, Schuller ER, Remick DG (2010) Early enhanced local neutrophil recruitment in peritonitis-induced sepsis improves bacterial clearance and survival. J Immunol 185:6930–6938
77. Osuchowski MF, Welch K, Siddiqui J et al (2006) Circulating cytokine/inhibitor profiles reshape the understanding of the SIRS/CARS continuum in sepsis and predict mortality. J Immunol 177:1967–1974
78. Osuchowski MF, Connett J, Welch K et al (2009) Stratification is the key: inflammatory biomarkers accurately direct immunomodulatory therapy in experimental sepsis*. Crit Care Med 37(5):1567–1573
79. Seymour CW, Kerti SJ, Lewis AJ et al (2019) Murine sepsis phenotypes and differential treatment effects in a randomized trial of prompt antibiotics and fluids. Crit Care 23:384
80. Turnbull IR, Javadi P, Buchman TG et al (2004) Antibiotics improve survival in sepsis independent of injury severity but do not change mortality in mice with markedly elevated interleukin 6 levels. Shock 21:121–125
81. Yang Y, Xie J, Guo F et al (2016) Combination of C-reactive protein, procalcitonin and sepsis-related organ failure score for the diagnosis of sepsis in critical patients. Ann Intensive Care 6:51

Chapter 17

Detection of Blood Cell Surface Biomarkers in Septic Mice

Dinesh G. Goswami and Wendy E. Walker

Abstract

Sepsis arises when an infection induces a dysregulated immune response, resulting in organ damage. New methods are urgently needed to diagnose patients in the early stages of sepsis, and identify patients with a poor disease prognosis. One promising approach is to identify the rapid changes in cell surface antigens (biomarkers) that occur during sepsis, as a consequence of leukocyte mobilization and activation. This chapter describes the method for staining whole blood with fluorescently conjugated antibodies that detect cell surface biomarkers, and performing flow cytometry analysis to quantify biomarker-positive cells. Our protocol is designed to detect blood cell surface biomarkers in septic mice, but could also be applied to study potential biomarkers in blood obtained from human patients with sepsis and other medical conditions.

Key words Fluorescence activated cell sorting, FACS, Cecal ligation and puncture, CLP, Sepsis, Blood staining, Biomarker

1 Introduction

Biomarker analyses in biological fluids (blood, serum, urine, etc.) provides a minimally invasive method to diagnose disease conditions and pathologies. Quantification of biomarkers has been explored as a tool to predict, diagnose, monitor, and treat many diseases, including brain injury, cancer, and Alzheimer's disease [1–4]. Circulating leukocytes are a major component of the immune system. Neutrophils and monocytes rapidly infiltrate sites of injury, inflammation, and infection [5, 6]. Sepsis alters the leukocyte populations present in the bloodstream, as well as activating leukocytes. A state of dysregulated immunity ensues, encompassing persistent inflammation and immune suppression [7–10]. As a consequence, multiple cell surface markers show differential expression during sepsis. Therefore, measuring these cell surface biomarkers is a potential strategy to diagnose sepsis. Amongst the cell surface biomarkers that have been explored for sepsis diagnosis, promising results have been achieved with CD69, CD64, and CD25 [11–25].

Wendy E. Walker (ed.), *Sepsis: Methods and Protocols*, Methods in Molecular Biology, vol. 2321,
https://doi.org/10.1007/978-1-0716-1488-4_17,

Flow cytometry is powerful tool to analyze blood cells. Flow cytometers can rapidly analyze a large number of cells, distinguish single cells from doublets, and detect specific and rare cell types within a heterogeneous cell population. Apart from detecting the presence of particular cell populations, the state of activation of the cells can also be assessed [26]. The scope of flow cytometry is ever expanding as we gain an increased understanding of how cell surface markers relate to basic biology, and due to the development of new antibodies. This makes flow cytometry a highly valuable tool for biomarker discovery, clinical validation, and routine implementation. Antibody staining can be conducted on whole blood drawn into heparinized or EDTA tubes, providing a simple and straightforward assay. Staining and flow cytometry analysis can be completed within 2 h, providing a rapid turnaround. However, staining the blood for biomarkers has its own challenges. Storage of blood as well as sample processing could affect biomarker expression analyses, and changes in absolute cell counts may confound data interpretation [27]. Therefore, standardization of the assay to stain for blood biomarkers is critical. Careful gating, and inclusion of appropriate controls are also important to ensure the validity and scientific rigor of the study.

In this chapter we describe a simple, rapid protocol for staining whole blood cells with antibodies, and present an example using the CD69, CD64, and CD25 biomarkers. Although this protocol uses mouse blood, the procedure is compatible with human blood (substituting human antibodies and Fc blocking reagent) and could be used for clinical research on sepsis and other medical conditions.

2 Materials

2.1 Staining Whole Blood

1. Potassium-EDTA tubes for blood collection.
2. Staining buffer: 2% FBS in PBS (ice cold, use within 1 week of preparation).
3. Optional: mouse Fc blocking reagent (*see* **Note 1**).
4. Fluorescently labeled antibodies against the cell surface markers of interest.
5. 1 × BD FACS Lysing solution (Becton Dickinson): dilute 1 volume of 10 × BD FACS Lysing solution with 9 volumes of dH_2O.
6. 20 μL, 200 μL, and 1 mL pipetmans.
7. Pipette tips.
8. Ice bucket, filled with ice.
9. 1.5 mL microcentrifuge tubes.

10. Benchtop microcentrifuge.
11. Vortex mixer.
12. Biosafety cabinet.

2.2 Staining Compensation Beads (Optional)

1. Compensation beads (ensure you select beads that bind to the antibodies used).
2. Staining buffer (*see* Subheading 2.1).
3. Fluorescently labeled antibodies against the cell surface markers of interest.
4. Isotype controls for each antibody (labelled with the same fluorophore).
5. 20 μL, 200 μL, and 1 mL pipetmans.
6. Pipette tips.
7. Ice bucket, filled with ice.
8. 1.5 mL microcentrifuge tubes.
9. Benchtop microcentrifuge.
10. Vortex mixer.

2.3 Flow Cytometry and Analysis

1. Flow cytometer.
2. 5 mL round bottom polystyrene tubes.
3. Transfer pipette, or 1 mL pipetman and tips.
4. 10% bleach in dH_2O.
5. Optional: 5% Contrad in dH_2O.
6. dH_2O.
7. FACS analysis software.

3 Methods

3.1 Staining Blood

1. Prepare a scheme for antibody staining, detailing the sample and antibodies that will be added to each tube. Appropriate controls should be included, such as unstained, single stain controls, isotype controls and fluorescence minus one (FMO) controls to account for autofluorescence, specificity, and compensation (*see* **Note 2**). An example is provided in Table 1 and the purpose of the various controls are discussed in Table 2.
2. Prepare a working cocktail of antibodies at the appropriate dilutions in ice-cold staining buffer. The working solution should contain antibodies at 10× the final concentration (*see* **Note 3**). Make enough volume to add 5 μL per tube, plus an extra 5 μL to account for loss during pipetting. Prepare FMO working solutions in the same manner, replacing the relevant antibody with staining buffer. Store these on ice until use.

Table 1
Sample staining scheme

	Sample	Fc block	CD69	CD64	CD25	Isotype (CD69)	Isotype (CD64)	Isotype (CD25)
1	Unstained	+						
2	Single stain—CD69	+	+					
3	Single stain—CD64	+		+				
4	Single stain—CD25	+			+			
5	Isotype for CD69	+				+		
6	Isotype for CD64	+					+	
7	Isotype for CD25	+						+
8	CD69 FMO	+		+	+			
9	CD64 FMO	+	+		+			
10	CD25 FMO	+	+	+				
11	Fully stained sample (add additional tubes for each blood sample)	+	+	+	+			

Table 2
Staining controls/reagents and their purpose

Sample	Purpose
Unstained control	Negative control: Defines the background autofluorescence signal
Single stain controls	Positive controls: Define the specific signal from each individual antibody. These can be used to perform and verify compensation
Isotype stained controls	Negative control: Isotype antibody (conjugated to the same fluorophore as the antibody used in the assay) can help to distinguish nonspecific antibody binding from true antibody staining
Fluorescence minus one (FMO) controls	Negative control: Helps to define the background signal for a particular antibody and to set the gate for positive cells
Compensation beads	These can be used for compensation in place of the unstained and single stained controls
Fc block	Blocks nonspecific binding of antibodies to the Fc receptor. This reagent can be very important to prevent nonspecific binding, but may inhibit true binding of antibodies that target the Fc receptor

Example: to stain 9 samples with anti-CD69, anti-CD64, and anti-CD25 (each used at 1/200 final dilution), prepare the following working cocktail:

anti-CD69 (1/20): 2.5 μL.

anti-CD69 (1/20): 2.5 µL.

anti-CD69 (1/20): 2.5 µL.

staining buffer: 42.5 µL.

final volume: 50 µL.

3. For each antibody, prepare a working single stain solution for a single stain control at 10× the final concentration (*see* **Note 3**). Make 10 µL final volume. Prepare an isotype control for each antibody, in the same manner. Store on ice until use.

   Example: To make a working single stain solution of anti-CD69 (used at1/200 final dilution):

   anti-CD69 (1/20): 0.5 µL.

   staining buffer: 9.5 µL.

   final volume: 10 µL.

4. Obtain blood samples from septic mice and appropriate controls in K2-EDTA tubes (*see* **Notes 4–7**).
5. Label a set of 1.5 mL microcentrifuge tubes, according to the staining scheme. Transfer 44 µL of the whole blood into the appropriate tubes (*see* **Note 8**). For the single stain and isotype controls, select a blood sample that is expected to show a positive signal.
6. Optionally, add 1 µL mouse Fc Blocking reagent to each tube, vortex gently to mix, and incubate at room temperature for 5 min (*see* **Note 1**).
7. Add 5 µL of the working cocktail of antibodies to the appropriate tubes, and 5 µL of the working single stain, isotype control, and FMO solutions to the appropriate tubes. Add 5 µL staining buffer to the unstained control. Vortex each tube gently to mix.
8. Incubate for 30 min in the dark at room temperature (*see* **Note 9**).
9. After the incubation, add 1 mL of 1× FACS Lysing solution to each sample, to lyse RBC and fix the cells (*see* **Note 10**).
10. Vortex gently and incubate for 10 min in the dark at room temperature.
11. Centrifuge at 500 × *g* for 5 min. Remove as much supernatant as possible, while retaining the cell pellet.
12. Add 1 mL staining buffer to wash the cells, vortex gently to mix, and centrifuge at 500 × *g* for 5 min. Remove as much supernatant as possible while retaining the cell pellet.
13. Add 250 µL of staining buffer to the pellet and vortex gently to resuspend.

3.2 Staining Compensation Beads

Compensation is required for experiments that measure more than one biomarker per sample, by staining the sample with multiple antibodies conjugated to different fluorophores. When these fluorophores have some overlap in terms of their emissions spectra (e.g., FITC and PE), it is necessary to apply signal compensation to detect the individual biomarkers. To calculate this compensation, you will need an unstained sample and single-stain samples that express a good positive signal for each antibody. In some cases, the single-stain blood samples can serve as compensation controls. However, it is often preferable to use compensation beads for this purpose, especially when one or more biomarkers are weakly expressed on blood leukocytes. The compensation beads include a population of beads that binds to the antibody avidly (these beads are coated with antibodies recognizing species-specific antibody light chains), and a second population of beads that do not bind to the antibody. When selecting compensation beads, ensure that they bind to the antibodies used (based on the species wherein the antibody was raised, and the isotype of the antibody). Here we present a protocol to stain beads for compensation controls.

1. Label a set of microcentrifuge tubes—one for the unstained bead control and one for each fluorophore.
2. Aliquot 10 μL of beads to each tube.
3. Add 35 μL staining buffer to each tube.
4. Prepare working single stain antibody solutions at 10× the final concentration, as described in Subheading 3.1, **step 3** (if solution remains from that preparation, it can be reused for the bead staining, but it should be used within 1 day of preparation).
5. Add 5 μL of the working single stain antibody solutions to the appropriate tubes and vortex briefly to mix.
6. Incubate for 15–30 min on ice in the dark.
7. Wash with 500 μL staining buffer and spin at 500 × *g* for 2 min. Remove as much liquid as possible while retaining the bead pellet.
8. Resuspend in 250 μL staining buffer and vortex to resuspend.

3.3 Flow Cytometry

1. Vortex the blood and bead samples gently and transfer them into 5 mL round-bottom polystyrene tubes.
2. Warm up the flow cytometer and start up the fluidics.
3. Ensure that the machine has been calibrated and that it has passed a daily performance check (follow the instructions for your flow cytometer).
4. Set up the experiment in the acquisition software. Set up the machine to read FSC-A, FSC-H, SSC-A, and the fluorophores

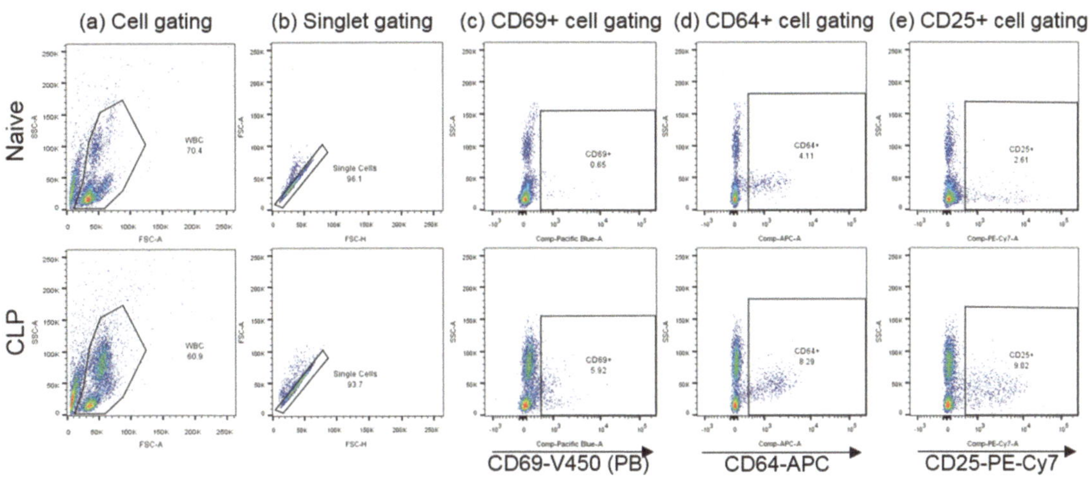

Fig. 1 CD69, CD64, and CD25 biomarker staining in whole blood from naïve and CLP (18h post-surgery) mice. The mouse that has undergone CLP has more cells positive for CD69, CD64, and CD25. Whole blood was stained with V450 anti-mouse CD69, APC anti-mouse CD64, and PE-Cy7 anti-mouse CD25 antibodies. Representative plots for (**a**) cell gating, (**b**) singlet gating, and positive gates for (**c**) CD69, (**d**) CD64, and (**e**) CD25 are shown here

used in the experiment. FSC and SSC should be read with a linear scale, while fluorophores are read with a logarithmic scale.

5. Adjust the FSC and SSC amplitude to yield good separation of the leukocyte populations (Fig. 1a). Adjust the amplitude for each fluorophore to achieve a good separation of the positive and negative cells (Fig. 1c–e, *see* **Note 11**).
6. Create compensation controls for the unstained sample and single stained samples. Run these samples through the machine and record the data. For this purpose, it will typically suffice to collect 1000 events. Adjust the event gate to collect the beads (based on FSC vs. SSC scatter) and adjust the compensation gates to segregate the positive and negative populations for each fluorophore. Instruct the machine to calculate the compensation (*see* **Note 12**).
7. Run the samples through the flow cytometer and record the data (vortex each tube gently immediately before use). Typically, 10,000 events are collected for each sample. However, a larger number of events (e.g., 50,000) may be collected when looking for rare events, and a smaller number of events may be collected (e.g., 1000) when sample material is limited.
8. Export the data files in a format that can be read by the analysis software (fcs format for FlowJo).
9. Clean and disinfect the flow cytometer according to the manufacturer's instructions. Typically, this may constitute running 10% bleach, 5% Contrad and then dH_2O through the machine for 5 min each. Shut down the fluidics. Switch off the machine.

3.4 Analysis of FACS Data

1. Open the analysis software and load the flow cytometry files.
2. Open the first sample, and create graph of FSC-A versus SSC-A (Fig. 1a).
3. Draw a gate around the leukocyte populations of interest (*see* Fig. 1a and **Note 13**).
4. Apply this gate to all samples (*see* **Note 14**).
5. Open the gated leukocytes for one sample and create a graph of the FSC-H vs FSC-A (Fig. 1b). Draw a gate around the singlets. Note that cells with a high FSC-A:FSC-H ratio are doublets (two cells stuck together), these are typically excluded from the analysis.
6. Apply this gate to all samples (*see* **Note 14**).
7. Decide if you will gate positive versus negative cells or compare the mean or median fluorescence intensity of the biomarker. The former is appropriate when sepsis induces a clear, separate population, the latter is appropriate when the entire population shifts up the scale (mean or median may be selected depending on the data distribution).
8. To gate a positive cell population: Within the singlets, create a graph for each fluorophore. Draw a box around the cells positive for each biomarker (according to the fluorophore that was conjugated to that antibody). This will reveal the proportion of cells that are positive for each biomarker (Fig. 1c–e). Apply this gate to all samples. Compare the single stain controls to the unstained and isotype-stained controls. The former should show a signal while the latter two do not. Likewise, compare the fully stained sample to the FMO control (omitting the antibody of interest), the former should show signal while the later does not. Adjust the gating if necessary (*see* **Note 15**).
9. Alternatively, instruct the software to calculate the mean or median fluorescence intensity for the fluorophore of interest. Apply the statistic to all samples.

3.5 Expected Results

Figure 1 shows an example FACS plots for CD69, CD64, and CD25 biomarker staining. This study was approved by the IACUC of TTUHSC at El Paso. One female C57BL/6J mouse was subject to CLP, per our prior reports [28–32] and administered Lactated Ringer's Solution (1 mL s.c.) postop and primaxin (25 mg/kg s.c.) at 3 h post-surgery. A blood sample was obtained at 18 h post-CLP surgery and from a naïve female C57BL/6J mouse. An increase in the percentage of biomarker positive cells can be observed in the CLP mouse, compared to the naïve mouse (Fig. 1c–e). Please note that our prior report found that CD69 and CD64 are valuable biomarkers to rapidly detect sepsis in mice [32]. However, CD64 and CD25 were also upregulated in mice subject to sham surgery, during the early postoperative period [32]. This study demonstrated that mouse models are a useful tool to study and validate sepsis biomarkers [32].

3.6 Troubleshooting

Although this protocol is simple, there are several steps that may go wrong and potential issues that may confound the analysis. Table 3 shows a guide to troubleshooting the experiment and data analysis.

4 Notes

1. Fc Blocking reagent will reduce nonspecific binding of antibodies to cells expressing Fc receptors (such as B cells and monocytes). However, this reagent could potentially reduce true staining with antibodies that detect Fc receptors (such as CD64). In the protocol presented here, we observe little difference in staining signal between samples treated with Fc block and untreated samples. However, we have found that for other antibodies, it is critical to include Fc block to obtain valid results. An isotype control can help to differentiate true staining from nonspecific binding, and the protocol can be tested in the presence and absence of Fc Blocking reagent to determine if it should be used.
2. Once the investigator has established the protocol, it may not be necessary to run all of these controls in subsequent experimental repeats, provided that the same settings are used for the flow cytometer, and the machine is calibrated on a regular basis.
3. Each antibody should be titrated to determine the appropriate concentration/dilution of the antibody for optimal staining. This can be accomplished by making serial dilutions of the antibody and testing for staining in cells known to express the relevant cell surface antigen/marker. The investigator is advised to read the product insert, or contact the manufacturer for more information on each specific antibody. The antibody solutions can be prepared 1 day in advance of the staining procedure.
4. Blood samples may be collected from mice by piercing the submandibular vein with a lancet, or from the retro-orbital sinus using a heparinized capillary tube. The latter technique has been described in detail in our prior methods chapter [33]. Although the submandibular route may be preferred by some IACUCs, the retro-orbital technique may allow better control of the blood collection volume. When performed by a proficient investigator with the appropriate anesthesia (isoflurane and proparacaine eye drops), the retro-orbital bleed will not cause much discomfort to the animals. Serial bleeds can be performed on an animal, provided the sum of all bleeds within a

Table 3
Troubleshooting biomarker staining and FACS analysis

Issue	Potential explanation	Solutions
No signal is observed with the antibody	• The biomarker is not expressed on the cells in the blood sample • An inappropriate antibody dilution was used • The antibody reagent is bad • The voltage level for that fluorophore has been set too low on the flow cytometer • A technical error was made when staining (no antibody, or incorrect antibody was added)	• Test and titrate the antibody using a cell population that is known to express the antigen • Test the antibody using compensation beads. If no signal is observed here, the antibody reagent is likely bad • Adjust the voltage for that fluorophore on the flow cytometer • Repeat the stain to address the possibility of technical error
Signal is observed in the isotype control	• Nonspecific binding by Fc-expressing cells or dead cells	• Use Fc block to reduce nonspecific binding by cells expressing the Fc receptor • Use a viability dye to exclude dead cells (ensure the reagent is compatible with fixing cells)
A single stain sample yields signal in two fluorescent channels	• The fluorescence was not correctly compensated • The fluorophores are incompatible • Degradation of tandem fluorophore dyes to the parent dye (e.g., PE-Cy7 to PE)	• Adjust the compensation • If needed, select different fluorophores or measure the two biomarkers in separate tubes • Use a stabilizing fixative to preserve tandem dyes, or select a nontandem dye instead
A single stain sample causes "spreading," for example, widening of the background signal in a second channel	• This is an artifact of compensation from certain fluorophores (e.g., PE and PERCP-Cy5.5)	• Select different fluorophores for one or both antibodies, or stain the biomarkers in separate tubes
The machine is unable to calculate the compensation	• The sample yields inadequate signal to calculate compensation. Either the antibody is bad or the cells do not express the antigen • A technical error was made when staining the compensation controls (no antibody, or incorrect antibody was added)	• Stain compensation beads. Ensure they bind to the antibody subtype being used. If no signal is observed in flow cytometric analysis, the antibody reagent is probably bad. If signal is observed, these beads can be used for the compensation setup • Repeat the stain to address the possibility of technical error
No events are observed when the sample is run through the flow cytometer	• The cells have settled to the bottom of the tube • The machine is clogged	• Vortex the sample gently immediately before running it through the flow cytometer to resuspend the cells

(continued)

Table 3 (continued)

Issue	Potential explanation	Solutions
	• The cells have been lost during processing	• Test the machine by running compensation beads. If no events are observed, the machine may be clogged. Follow the manufacturer's instructions to unclog the machine. Good maintenance (including cleaning the machine after each use) will minimize clogging • If the beads show signal, but the sample does not, the cells may have been lost (you can test this by applying the sample to a hemacytometer slide). Test the protocol again, taking care not to disrupt and suck up the cell pellet when washing (remove the supernatant quickly after the centrifuge step and leave a small volume of buffer at the bottom to avoid disrupting the cell pellet) • Additionally, make sure not to lyse the cells for too long, this can lead to an insoluble cell pellet
An insoluble clump is observed in the staining tube	• The cells have been damaged during the lysis/fixation step • Cell to cell adhesion has caused a clump • Debris was present in the tube	• Do not run this sample through the flow cytometer as it may clog the machine • Shorten the lysis/fixation step • Add EDTA to the FACS buffer to prevent cell clumping • Ensure clean tubes are used
Only very small events are observed when running the sample through the flow cytometer	• The FSC:SSC voltages are set too low • The cells have been destroyed during processing and only debris is present • The staining buffer is contaminated with bacteria • The mouse has a very low white blood cell count	• Adjust the FSC:SSC setting • Apply the sample to a hemacytometer slide to determine if WBC are present in the sample, before and after the staining protocol. If the cells are destroyed during the processing, you may need to adjust your handling of the sample. Take care not to vortex the cells too vigorously, do not spin at a high g-force and do not lyse the blood for too long

(continued)

Table 3
(continued)

Issue	Potential explanation	Solutions
		• Test the staining buffer for cloudiness, and read it on the flow cytometer to see if bacteria are present in the buffer itself (they will read as small events). If needed, replace the buffer • We have observed that some mice that are extremely ill from sepsis exhibit extreme leukopenia, making it difficult to read cell surface biomarker expression levels
The investigator is unsure how to set the gate for positive cells	• Controls can be used to address this issue	• Compare the fully stained sample to the FMO sample, and the unstained sample to the single stain sample, to determine where to set the gate

2-week period is ≤10% of the total blood volume or ~1% of the body weight of the mouse. It is recommended to alternate between the left and right eye for serial bleeds.

5. Blood samples are potentially infectious and therefore appropriate personal protective equipment should be donned and precaution should be taken while handling them. Blood samples should be disposed according to relevant regulations/ guidelines at your institution. While this risk is minimal when using mouse models of sepsis derived from commensal bacteria (such as CLP), enhanced precautions may be required for other pathogens, and when human blood is used. Consult your institutional biosafety committee for guidance.

6. Other anticoagulants may be more advantageous in particular cases. We have only used K2-EDTA but Na-Heparin was shown to better preserve cell surface marker expression in a prior report [27]. However, the total cell count may show a pronounced decrease with Na-Heparin. Therefore, it is important to be consistent with the use of anticoagulant when comparing samples.

7. Storage of blood before staining could lead to a change in the proportions and characteristics of the cell populations. Different cell populations in the blood have different half-lives. Additionally, the duration and temperature of blood storage may

induce biochemical changes (such as altered pH, blood gases, amino acids, carbohydrates, and lipids) that influence cell viability. Therefore, both time and temperature of blood storage need to be optimized for each study and should remain consistent between samples.

8. The assay could be scaled down for smaller blood volumes, or scaled up to look for rare events. We have used blood volumes ranging from 22.5 μL to 90 μL for staining without experiencing any issues. All other volumes and reagents should be scaled accordingly.
9. Longer incubation periods may be required in some cases, particularly for antigens that are expressed at low levels.
10. This protocol uses BD FACS lysing solution. Other lysis reagents may be used, according to the manufacturer's instructions. This protocol may not work well with certain hematologic disorders where RBCs are difficult to lyse (e.g., myelofibrosis, sickle-cell anemia, thalassemia, and spherocytosis). Incomplete lysis may occur if the blood is not mixed properly with the anticoagulants at the time of blood draw, if the blood is stored for an extended period or if the reagents are improperly mixed.
11. To determine the optimal settings, you can "acquire" a small number of events from one or more stained samples, but do not "record" them yet.
12. Using the BD FACSDiva software, the machine can be instructed to automatically calculate the optimal compensation settings. With other software, the compensation may need to be adjusted manually. In this case, create a graph for each pair of channels that exhibit emissions overlap. Gradually increase the percentage compensation until each fluorophore yields a signal for that fluorophore only, while the signal in the opposite channel is equivalent to the background level. Do not overcompensate.
13. In the example shown in Fig. 1, we have drawn the gate to include all the major blood leukocyte populations. However, it is possible to gate specific cell types (e.g., lymphocytes, monocytes, and neutrophils) according to their FSC:SSC scatter; our prior paper shows an example [32]. Note that events with a very low FSC and SSC profile represent platelets and erythrocytes, these are typically excluded from the analysis.
14. Look at all of the samples to ensure the gating is valid, and adjust if necessary. Ideally, the cell gate and the singlet gate should be identical for all samples.
15. The gate should be set such that the unstained, isotype control and FMO control for that fluorophore have no positive cells

within the gate. The single stain samples should yield a signal for the appropriate biomarker, and no signal for the other biomarkers. Ideally, the gate for each biomarker should be identical for all samples.

Acknowledgments

This study was supported by a grant from the Society of Critical Care Medicine (SCCM) and startup funds from Texas Tech University Health Sciences Center at El Paso, awarded to Wendy Walker.

References

1. Rohlwink UK, Figaji AA (2014) Biomarkers of brain injury in cerebral infections. Clin Chem 60:823–834
2. Van Giau V, An SS (2016) Emergence of exosomal miRNAs as a diagnostic biomarker for Alzheimer's disease. J Neurol Sci 360:141–152
3. Marrugo-Ramirez J, Mir M, Samitier J (2018) Blood-based cancer biomarkers in liquid biopsy: a promising non-invasive alternative to tissue biopsy. Int J Mol Sci 19:2877
4. An M, Gao Y (2015) Urinary biomarkers of brain diseases. Genomics Proteomics Bioinformatics 13:345–354
5. Kolaczkowska E, Kubes P (2013) Neutrophil recruitment and function in health and inflammation. Nat Rev Immunol 13:159–175
6. Shi C, Pamer EG (2011) Monocyte recruitment during infection and inflammation. Nat Rev Immunol 11:762–774
7. Delano MJ, Ward PA (2016) The immune system's role in sepsis progression, resolution, and long-term outcome. Immunol Rev 274:330–353
8. Ayala A, Chaudry IH (1996) Immune dysfunction in murine polymicrobial sepsis: mediators, macrophages, lymphocytes and apoptosis. Shock 6(Suppl 1):S27–S38
9. Gentile LF, Cuenca AG, Efron PA et al (2012) Persistent inflammation and immunosuppression: a common syndrome and new horizon for surgical intensive care. J Trauma Acute Care Surg 72:1491–1501
10. Ward NS, Casserly B, Ayala A (2008) The compensatory anti-inflammatory response syndrome (CARS) in critically ill patients. Clin Chest Med 29:617–625, viii
11. Cid J, Aguinaco R, Sanchez R et al (2010) Neutrophil CD64 expression as marker of bacterial infection: a systematic review and meta-analysis. J Infect 60:313–319
12. de Jong E, de Lange DW, Beishuizen A et al (2016) Neutrophil CD64 expression as a longitudinal biomarker for severe disease and acute infection in critically ill patients. Int J Lab Hematol 38:576–584
13. Dimoula A, Pradier O, Kassengera Z et al (2014) Serial determinations of neutrophil CD64 expression for the diagnosis and monitoring of sepsis in critically ill patients. Clin Infect Dis 58:820–829
14. Lam HS, Cheung HM, Poon TC et al (2013) Neutrophil CD64 for daily surveillance of systemic infection and necrotizing enterocolitis in preterm infants. Clin Chem 59:1753–1760
15. Rogina P, Stubljar D, Lejko-Zupanc T et al (2015) Expression of CD64 on neutrophils (CD64 index): diagnostic accuracy of CD64 index to predict sepsis in critically ill patients. Clin Chem Lab Med 53:e89–e91
16. Hirsh M, Mahamid E, Bashenko Y et al (2001) Overexpression of the high-affinity Fcgamma receptor (CD64) is associated with leukocyte dysfunction in sepsis. Shock 16:102–108
17. Leng FY, Liu JL, Liu ZJ et al (2013) Increased proportion of CD4(+)CD25(+)Foxp3(+) regulatory T cells during early-stage sepsis in ICU patients. J Microbiol Immunol Infect 46:338–344
18. Siqueira-Batista R, Gomes AP, Azevedo SF et al (2012) CD4+CD25+ T lymphocytes and regulation of the immune system: perspectives for a pathophysiological understanding of sepsis. Rev Bras Ter Intensiva 24:294–301
19. Chen K, Zhou QX, Shan HW et al (2015) Prognostic value of CD4(+)CD25(+) Tregs as

a valuable biomarker for patients with sepsis in ICU. World J Emerg Med 6:40–43
20. Boomer JS, Shuherk-Shaffer J, Hotchkiss RS et al (2012) A prospective analysis of lymphocyte phenotype and function over the course of acute sepsis. Crit Care 16:R112
21. Schwulst SJ, Muenzer JT, Chang KC et al (2008) Lymphocyte phenotyping to distinguish septic from nonseptic critical illness. J Am Coll Surg 206:335–342
22. Zhang Y, Li W, Zhou Y et al (2017) Detection of sepsis in patient blood samples using CD64 expression in a microfluidic cell separation device. Analyst 143:241–249
23. Zhang Y, Zhou Y, Li W et al (2018) Multiparameter affinity microchip for early sepsis diagnosis based on CD64 and CD69 expression and cell capture. Anal Chem 90:7204–7211
24. Zhou Y, Zhang Y, Johnson A et al (2019) Combined CD25, CD64, and CD69 biomarker panel for flow cytometry diagnosis of sepsis. Talanta 191:216–221
25. Zhou Y, Zhang Y, Johnson A et al (2019) Detection of culture-negative sepsis in clinical blood samples using a microfluidic assay for combined CD64 and CD69 cell capture. Anal Chim Acta 1062:110–117
26. Spijkerman R, Hesselink L, Hellebrekers P et al (2019) Automated flow cytometry enables high performance point-of-care analysis of leukocyte phenotypes. J Immunol Methods 474:112646
27. Diks AM, Bonroy C, Teodosio C et al (2019) Impact of blood storage and sample handling on quality of high dimensional flow cytometric data in multicenter clinical research. J Immunol Methods 475:112616
28. Heipertz EL, Harper J, Goswami DG et al (2021) IRF3 signaling within the mouse stroma influences sepsis pathogenesis. J Immunol 206(2):398–409
29. Heipertz EL, Harper J, Lopez CA et al (2018) Circadian rhythms influence the severity of sepsis in mice via a TLR2-dependent, leukocyte-intrinsic mechanism. J Immunol 201 (1):193–201
30. Heipertz EL, Harper J, Walker WE (2017) STING and TRIF contribute to mouse sepsis, depending on severity of the disease model. Shock 47:621–631
31. Walker WE, Bozzi AT, Goldstein DR (2012) IRF3 contributes to sepsis pathogenesis in the mouse cecal ligation and puncture model. J Leukoc Biol 92:1261–1268
32. Goswami DG, Garcia LF, Dodoo C et al (2021) Evaluating the timeliness and specificity of CD69, CD64 and CD25 as biomarkers of sepsis in MICE. Shock 55:507–518
33. Walker WE (2018) Methods to study the innate immune response to sepsis. Tharakan B in: traumatic and ischemic injury: methods and protocols. Springer New York, New York, NY, pp 189–206

Chapter 18

Microfluidic Chips for Sepsis Diagnosis

Yun Zhou, Yijia Yang, and Dimitri Pappas

Abstract

This chapter discusses two microfluidic-based approaches for early sepsis detection that achieve a higher accuracy than traditional blood culture analysis. Patient blood samples were included in this work to validate the performance of our chips in diagnosing sepsis. The single-parameter chip demonstrated the increased accuracy if using CD64 as a biomarker for sepsis detection compared with C-reactive protein (CRP) and procalcitonin (PCT) when applied alone. In addition, a multiparameter chip measuring a combined panel of CD25, CD64, and CD69, and achieved a high accuracy with an Area Under the Receiver Operating Characteristic Curve (AUROC) of 0.978. The combined panel was also able to detect culture-negative patients and provided a faster diagnosis. Besides, microfluidics has advantages of less time consuming, easier to manufacture, less sample loading, less complex, and portable. Therefore, our approach is of great potential to become a bedside sepsis detection method.

Key words Microfluidic chip, Sepsis diagnosis, Biomarker

1 Introduction

Microfluidic technologies are emerging as powerful tools for point-of-care (POC) analysis especially in cell separations [1]. With the microfluidic boom in last two decades, physical-based separations in microfluidic devices occupy a critical role in cell isolation [2]. These approaches have the advantages of less sample loading, less analysis time, and better separation compared with benchtop methods such as centrifugal devices [3]. However, affinity separations in which affinity molecule pairs can be used to identify specific cell types are the most selective method mainly in cases where physical differences are too small to be exploited [2]. Any molecule, including antibodies [4, 5], aptamers [6], and peptides [7], that shows selective bindings to membrane-bound molecules on the cell surface can be used as affinity capture molecules in microfluidic devices [2].

Our lab has proposed both single- and multiparameter microfluidic chips for sepsis diagnosis based on affinity surface cell capture [8, 9] (Fig. 1). In this method, antibodies are coated on the

Wendy E. Walker (ed.), *Sepsis: Methods and Protocols*, Methods in Molecular Biology, vol. 2321,
https://doi.org/10.1007/978-1-0716-1488-4_18,

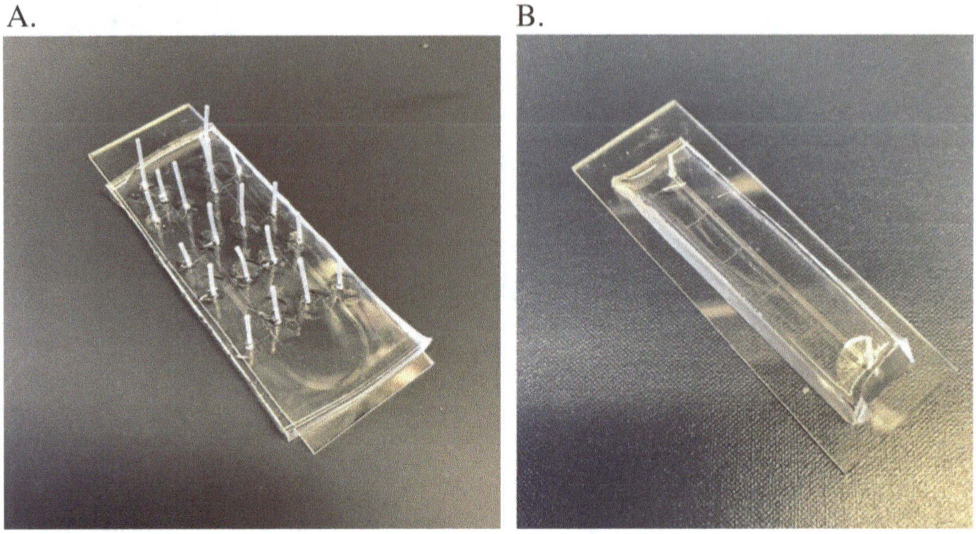

Fig. 1 (**a**) Picture of multiparameter microfluidic chip. The channel surface was coated with anti-CD25, anti-CD64, and anti-CD69 in different affinity regions for capturing target cells. (**b**) Picture of single-parameter herringbone microfluidic chip. The channel surface was modified with anti-CD64 for capturing mainly neutrophils

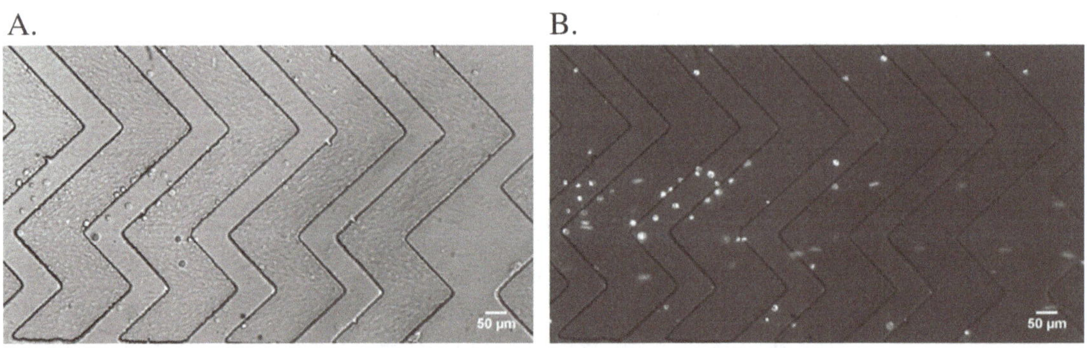

Fig. 2 Images of herringbone chip under (**a**) white light and (**b**) fluorescence light. White points in (**b**) are captured cells and blurry straight lines are moving cells

protein-modified channel surface via hydrogen binding to capture desired cells. Capture images are obtained under an inverted epi-fluorescence microscope (Fig. 2). The antibody retains target leukocytes while other background cells are eluted as waste. Many factors affect receptor-ligand bindings including receptor number, affinity between receptor and ligand, bond formation/dissociation rate, receptor diffusivity, distractive fluid forces, and contact areas [2]. Therefore, the selection of appropriate affinity ligand is extremely important to achieve the highest capture efficiency and capture purity. We have employed anti-CD64 in single-parameter chips, and anti-CD25, anti-CD64, and anti-CD69 in multiparameter chips for early sepsis detection. CD64, known as the high-affinity receptor to Fc fragment of IgG, has been reported to have a high surface expression on neutrophils and monocytes in sepsis patients [10]. Also, CD25 and CD69 were reported to be highly expressed on lymphocytes in sepsis [10]. Therefore, our multiparameter chip is able to obtain more comprehensive results compared with normal single parameter analysis. A detailed clinical study with human blood samples was conducted to validate these assays for sepsis diagnosis. We discovered that CD64 was a better biomarker for sepsis detection compared with other commonly used biomarkers such as C-reactive protein (CRP) and procalcitonin (PCT). Moreover, the multiparameter panel had 98% accuracy. We compared our results with clinical blood culture detection. These studies showed that our approach was capable of detecting not only culture-positive patients but also culture-negative patients [9].

2 Materials

All solutions should be prepared at room temperature. Storage of solutions and reagents is indicated separately below. Biohazardous waste should be processed in an autoclave before disposal. All biohazardous waste and normal waste should be disposed according to local waste disposal regulations.

2.1 Microfluidic Chip Fabrication

1. Channel mask: Microfluidic channels should be designed with Adobe Illustrator, or similar software, and then printed on a mask (20,000 dpi laser printer transparency).
2. 100 mm silicon wafers (University Wafer).
3. Negative photoresist SU-82025 (Micro Chem).
4. SU-82025 developer (Micro Chem).
5. 1H,1H,2H,2H-perfluorooctyltrichlorosilane (Alfa Aesar).
6. Polydimethylsiloxane (PDMS): SLYGARD™ 184 Silicone Elastomer Base.
7. PDMS Curing Agent: SLYGARD™ 184 Silicone Elastomer Curing Agent.
8. 30-gauge poly(tetrafluoroethylene) (PTFE) tubes.
9. Glass dish.
10. Weighing boats.
11. Scale.
12. Vacuum desiccator.
13. Oxygen plasma chamber.
14. Hot plate ×2.

15. Oven.
16. Fume hood.
17. Spin coater capable of 1358 × *g* with vacuum pump.
18. UV illuminator.
19. 100 mm quartz disc.

2.2 Microfluidic On-Chip Detection

1. Deionized water for blood lysis.
2. T50 Buffer: 10 mM Tris–HCl, 50 mM NaCl, pH 8.0; Prepare 500 mL deionized water in a graduated cylinder. Weigh 1.567 g Tris–HCl and 2.922 g NaCl, and then transfer to the cylinder. Add deionized water to the cylinder to make up to 1 L. Store at room temperature.
3. Biotinylated bovine serum albumin: Biotin-BSA, 1 mg/mL in T50 buffer. Obtain 5 mg Biotin-BSA. Dissolve the solid into 5 mL T50 buffer in a 10 mL centrifuge tube (*see* **Note 1**). Store in the freezer (−20 °C).
4. Neutravidin: Prepare 10 mL T50 buffer in a 10 mL centrifuge tube. Weigh 10 mg neutravidin and transfer into T50 buffer. 1.0 mg/mL solution must be diluted to 0.2 mg/mL with T50 buffer before use (*see* **Note 2**). Store in the freezer (−20 °C).
5. Phosphate-Buffered Saline: PBS, 1×, without calcium and magnesium; Corning Cellgro. Store at room temperature.
6. 3% Bovine Serum Albumin (BSA) in PBS buffer: Obtain 1.5 g Bovine Serum Albumin (Sigma-Aldrich) and dissolve in 50 mL PBS in a 50 mL centrifuge tube (*see* **Note 3**). Store at 4 °C. Warm up in the water bath to 37 °C before use.
7. 10× Concentrated Saline Buffer: 80 g/L 10× NaCl solution; Prepare 50 mL deionized water in a 50 mL centrifuge tube. Weigh 4 g NaCl and transfer to the deionized water. Store at room temperature.
8. Syringe, 1 mL.
9. PTFE tube 29 gauge.
10. Syringe pump.
11. Hemacytometer slide.
12. Microcentrifuge tubes.
13. Pipette and tips.
14. 300 mm OD fused silica capillaries, filled with solidified adhesive, to block the side channels (for multiparameter detection).
15. Water bath.
16. Fluorescence microscope, attached to a computer for image collection.

2.3 Antigens and Conjugates

1. Anti-human CD25 biotin, 0.5 mg/mL, clone BC96; Obtain 5 μL anti-CD25 and then dilute with 45 μL PBS for chip surface modification (1:9). Store at 4 °C.
2. Biotin mouse anti-human CD64, clone 10, 100 tests in 2 mL; Store at 4 °C.
3. Anti-human CD69 biotin, clone FN50, 0.5 mg/mL. Obtain 5 μL Anti-CD69 and then dilute with 45 μL PBS for chip surface modification (1:9). Store at 4 °C (*see* **Note 4**).
4. Hoechst 33342, trihydrochloride, trihydrate, 1 mg/mL.

3 Methods

3.1 Silicon Wafer Fabrication

1. Preheat the hot plate to 95 °C and the oven to 200 °C.
2. Turn the UV lamp on.
3. Gently remove dust from the wafer surface with air (*see* **Note 5**).
4. Place the wafer in the spin coater chamber.
5. Disperse negative photoresist SU-82025 (Micro Chem) on the silicon wafer surface (*see* **Note 6**).
6. Turn on the vacuum pump, nitrogen, and the spin coating sequentially. The silicon wafer surface is modified by 6 s 500 rpm and 30 s 1000 rpm spinning to form an evenly distributed 40 μm thick photoresist layer.
7. Turn off the nitrogen and vacuum pump.
8. Heat the wafer on a hot plate at 95 °C for 5 min.
9. Mount the wafer in the UV illumination system and place the mask on it (*see* **Note** 7). Place the quartz disc over the mask to hold it in place. Then place the whole setup in the UV chamber. Close the chamber door.
10. Turn the UV on for 10 s.
11. Heat the wafer on the hot plate for 5 min at 95 °C. Remove from heat. If the polymer has cured under UV, you will see a faint outline of the chip design at this point.
12. Under the fume hood, wash the wafer with SU-8 developer in the glass dish. To wash, place the wafer in the glass dish and pour just enough developer to cover the surface of the wafer. Swirl the dish allowing developer to wash away uncured polymer for 3 min. Discard the developer as hazardous waste.
13. Add just enough isopropyl alcohol to cover the wafer surface. Swirl the liquid to wash away any residual developer. The alcohol turns cloudy as it dissolves the developer. Discard the alcohol. Repeat **steps 11** and **12** one more time. The wafer

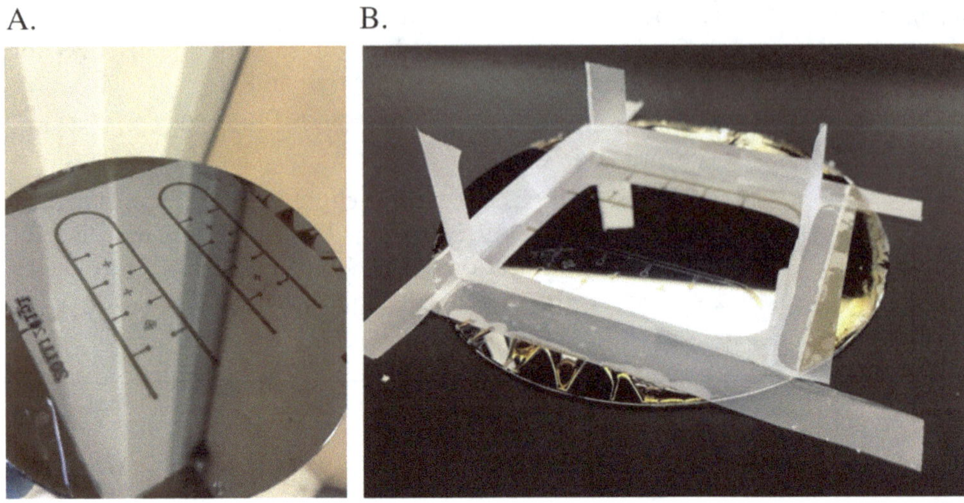

Fig. 3 (**a**) Wafer containing channel features after isopropyl alcohol washing. (**b**) Square dam region with low-tack tape on the wafer surface for PDMS mixture loading

should have a shiny, mirror-like surface, and the chip design will be fully developed at this point (Fig. 3a).

14. Heat the wafer for 10 min at 200 °C in an oven.
15. Place the wafer in the vacuum desiccator. Add 3–4 drops of 1H,1H,2H,2H-perfluorooctyltrichlorosilane to the vacuum desiccator and leave the wafer overnight (≥12 h) under vacuum to make the surface hydrophobic. The wafer is now ready for use. The wafer can be used multiple times, inspect the channels for signs of wear to determine when a new wafer is needed.

3.2 Single-Parameter Herringbone Chip Fabrication

1. Preheat the hot plate to 120 °C and the oven to 80 °C.
2. Weigh around 8 g PDMS in a weighing boat (*see* **Note 8**). Calculate the mass of PDMS curing agent. The ratio of PDMS and PDMS curing agent is 10:1 (*see* **Note 9**). Add the curing agent to the PDMS. Stir until completely mixed.
3. Place the weighing boat in the desiccator under vacuum for 20 min to remove bubbles.
4. Make a square dam region with low-tack tape on the wafer surface (include all channel features inside the square region) (Fig. 3b).
5. Pour the mixture into the tape square (*see* **Note 10**). Save the excess PDMS mixture for **step 13**. Place the wafer on the hot plate for at least 40 min at 120 °C.
6. Peel the PDMS from the wafer. Cut the PDMS slide into a 6 cm × 2 cm rectangle (include all channel features). Discard PDMS that does not have channel features.

7. Punch holes on the PDMS slide as inlets and outlets as indicated by channel features. Clean the slide surface with tape.
8. Place a glass slide and the PDMS slide into the oxygen plasma chamber. Turn on the vacuum pump for 3 min. Introduce oxygen into the chamber for 1 min 30 s. Warm up for 30 s and then adjust the intensity to high for 1 min (*see* **Note 11**).
9. Carefully attach the PDMS slide to the glass slide. Remove bubbles between these two layers. The PDMS and glass will bond irreversibly.
10. Place the setup in an oven set to 80 °C for 5 min.
11. Insert 2-cm length 30-gauge poly(tetrafluoroethylene) (PTFE) tubes into holes punched in **step 6** as inlets and outlets.
12. Place the setup in 80 °C oven for another 5 min.
13. Seal the bottom of all inlets and outlets with excess PDMS mixture left in the weighing boat (*see* **Note 12**). Keep the chip in the oven for 10 min. The chip is ready for use.

3.3 Multiparameter Double-Layer Chip Fabrication

1. The procedure for fabrication of the control (valve) layer is the same as the single-parameter herringbone chip. *See* Subheading 3.2 from **steps 1** to **6**. The ratio of PDMS: PDMS curing agent is 5:1.
2. Heat the hot plate to 70 °C.
3. For the channel layer: Weigh 5 g PDMS in a weighing boat. The ratio between PDMS and curing agent is 25:1. Therefore, calculate and add curing agent into PDMS (around 0.2 g). Stir the mixture until completely mixed.
4. Place the weighing boat in the desiccator for 20 min.
5. Place the fluidic layer wafer in the spin coater.
6. Pour the PDMS and curing agent mixture onto the middle of the wafer.
7. Turn the vacuum pump, nitrogen, and spin coating on sequentially. Refer to the manufacturer's instructions for your spin coater.
8. The fluidic channel is spin coated on the wafer surface at 2000 rpm for 30 s.
9. Place the wafer on a hot plate for exactly 30 min at 70 °C (*see* **Note 13**).
10. Heat the second hot plate to 120 °C.
11. Place the control layer on top of the fluidic layer and then heat the setup on a hot plate for at least 2 h at 120 °C (*see* **Note 14**).

12. Gently cut the PDMS slide (6 cm × 2 cm) down from the wafer.
13. Following the same **steps 7–13**, as outlined in Subheading 3.2.
14. The chip is ready for use.

3.4 Single-Parameter On-Chip Detection

1. Load enough deionized water into the channel to check for any leakages (*see* **Note 15**). Air-dry the channel before use (*see* **Note 16**).
2. Inject 8 μL biotin-BSA into the channel. Incubate for 45 min at room temperature.
3. Remove the biotin-BSA. Rinse the channel with T50 buffer.
4. Inject 8 μL Neutravidin into the channel. Incubate for 15 min at room temperature.
5. Remove the Neutravidin. Rinse the channel with T50 buffer and deionized water sequentially.
6. Air-dry the chip and store it at 4 °C (*see* **Note 17**).
7. Inject 8 μL anti-CD64 into the channel. Incubate for 20 min at room temperature.
8. Obtain a 100 μL blood sample. Warm it up to room temperature. Rotate to homogenize.
9. Blood lysis: Add 900 μL deionized water; mix for exactly 30 s; add 110 μL 10× NaCl buffer. Centrifuge at 1358 × *g* for 5 min. Discard supernatant.
10. Resuspend blood sample to a final volume of 1000 μL with PBS.
11. Add 1.4 μL Hoechst. Incubate in the dark for 20 min at room temperature.
12. Centrifuge the sample at 1358 × *g* for 5 min. Discard the supernatant.
13. Resuspend the sample in PBS to make a final volume of 1000 μL.
14. Repeat **steps 12** and **13** three more times to remove excess Hoechst and cell debris (*see* **Note 18**).
15. After the last wash, resuspend the sample in 3% BSA in PBS buffer to make a final volume of 100 μL (*see* **Note 19**). The sample is ready for the on-chip detection.
16. Connect the chip to a syringe containing 3% BSA in PBS buffer, using a 10 cm PTFE tube. Fill the chip with 3% BSA in PBS buffer. Turn on the white light lamp, fluorescent light lamp, computer software to control the microscope. Check the whole channel under the microscope and remove bubbles (*see* **Note 20**).

17. Draw 80–90 μL of lysed blood sample into a new syringe. Insert the sample syringe into the PTFE tube connected to the chip (*see* **Note 21**).
18. Place the sample syringe on a syringe pump. Set the flow rate to 0.6 mL/h and continue until the whole channel is filled with the blood sample. Stop the pump.
19. Check the whole channel under the microscope again. Make sure there are no (1) bubbles, (2) clusters, and (3) nonspecific binding.
20. Start the syringe pump at the flow rate of 0.06 mL/h. Wait for 20 min for continuous flow on-chip cell capture.
21. While waiting for the on-chip cell capture, put the hemacytometer slide under the microscope. Obtain a 10–20 μL blood sample and apply to the hemacytometer slide. Take an image to count the cells.
22. After 20 min cell capture, keep the flow rate at 0.06 mL/h. Start taking images of the whole channel under both white light and fluorescent light. The moving cells look like a straight line, while the captured cells are white points (Fig. 2b).
23. Stop the pump. Turn off the white light lamp and the fluorescent light lamp. Discard syringes in a sharps container. Discard the centrifuge vials, pipette tips, and all blood-contaminated waste as biohazardous waste.

3.5 Multiparameter On-Chip Detection

1. Pump air into the main control valve to block side channels (Fig. 4). Load enough deionized water into the main channel to check any leakages. Air-dry the channel before use (*see* **Notes 15** and **16**). Keep the air flow on until the completion of the first and second layer modification.
2. Inject 10 μL biotin-BSA into the main channel. Incubate for 45 min at room temperature.
3. Remove the biotin-BSA. Rinse the channel with T50 buffer.
4. Inject 10 μL Neutravidin into the main channel. Incubate for 15 min at room temperature.
5. Remove the Neutravidin. Rinse the channel with T50 buffer and deionized water sequentially. Stop the air flow.
6. Air-dry the chip and store it at 4 °C (*see* **Note 17**).
7. Pump air into the eight side control valves to separate the main channel into four regions (Fig. 4).
8. Inject 5 μL anti-CD64, anti-CD69, and anti-CD25 into three different regions as shown in Fig. 4. The last region is left for blank comparison. Incubate for 20 min at room temperature.
9. Blood sample preparation steps are the same as single-parameter on chip detection (*see* Subheading 3.4, **steps 8–15**).

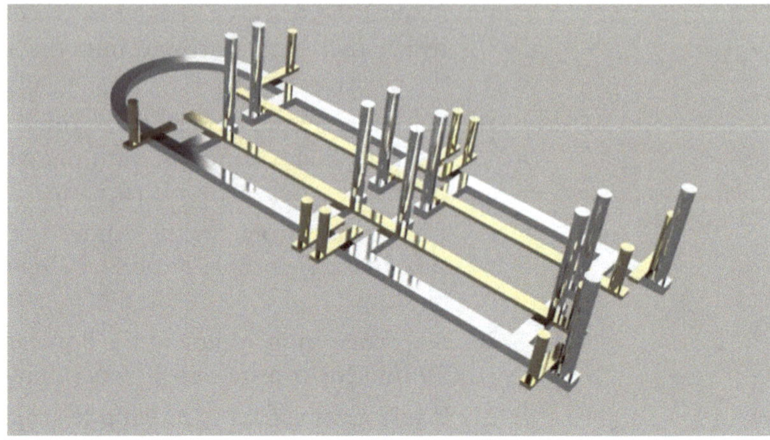

Fig. 4 Model simulation of multiparameter chip channel structure. This design has two separated layers. The bottom layer contains one main channel and eight side channels (white), and the top layer has two main valve and eight side valves (yellow). Main valves were used to control eight side channels for Biotin-BSA and neutravidin modification of the main channel. Eight side valves were used to separate the main channel into four different regions for different antibody surface modifications

10. When the chip and the blood sample are ready, use capillaries to block the side channels. The flow only goes through the main channel. Then, follow the same procedure described for single-parameter on chip detection (*see* Subheading 3.4, **steps 16–24**).

4 Notes

1. In order to achieve a more accurate concentration, you can prepare a larger batch and store it for future use. In this case, prepare 5 mL T50 buffer in a 10 mL centrifuge tube in advance. Add 5 mg Biotin-BSA and dissolve. Aliquot 500 μL of this solution into 10 microcentrifuge tubes. The vial that will be used immediately can be stored at 4 °C, while the other vials should be stored in a −20 °C freezer.
2. The concentration of the prepared solution is 1 mg/mL. For the same reasons described in **Note 1**, you can prepare a larger batch and store it for future use. For example, you can prepare 10 mL neutravidin solution, and aliquot 1 mL into 10 microcentrifuge tubes. As above, store the extra tubes in the freezer (−20 °C). Warm each tube to room temperature before use. Transfer the stored neutravidin solution to a 10 mL centrifuge tube and add 4 mL T50 buffer. The final concentration of the neutravidin solution should be 0.2 mg/mL. Again, aliquot

1000 μL solution into 5 microcentrifuge tubes. The vial that will be used immediately can be stored at 4 °C, the others should be stored in the freezer (−20 °C).

3. Bovine serum albumin (BSA) is hard to dissolve in PBS at room temperature, even with vortexing. Therefore, we usually place the centrifuge tube in a water bath at 37 °C for 20 min after adding the BSA the PBS. Subsequently, store the solution at 4 °C.
4. Anti-CD25 and anti-CD69 need to be diluted with PBS in a 1:9 ratio before loading into the chip. Anti-CD64 is ready for use at the purchased concentration.
5. The silicon wafer is photosensitive. Turn all lights off before use.
6. Check the vacuum index before spin coating. The index must be higher than 20 (or the value specified in the coater's instructions). If it is below 20, adjust the position of the silicon wafer to the middle of the wafer station. The SU-82025 must be dispersed in the middle of the wafer.
7. The mask contains channel features. Four masks were used in this work: fluidic layer of multiparameter chip, valve layer of multiparameter chip, channel layer of herringbone chip, and herringbone layer of herringbone chip. The fluidic layer and valve layer of the multiparameter chip need two separate wafers. They are all made with the same procedure described here. The channel of the herringbone chip is also made with this procedure. However, after finishing the channel layer, this wafer contains the straight channel which needs to be used again for the fabrication of the herringbone structure with herringbone mask. The fabrication of the herringbone layer also follows this procedure. The straight channel and the herringbone structure need to be exactly matched.
8. The mass of PDMS should be controlled at 8 ± 1 g. If the mass is more than 9 g, it would be difficult to punch holes in the following steps. Also, if the mass is less than 7 g, it would be difficult to insert inlets and outlets.
9. The PDMS and PDMS curing agent ratio needs to be strictly controlled at 10:1. Too much curing agent leads to softer PDMS. PDMS is hard to cure if too little curing agent was used.
10. Ensure that there are no bubbles in the mixture. If you cannot remove all bubbles inside the mixture on the first try, place the wafer containing the mixture into the desiccator again.
11. After finishing the oxygen plasma, it is necessary to open the vacuum chamber slowly to avoid dust attaching to the glass and PDMS slide.

12. The sealing step needs to be done in the oven and the oven temperature needs to be maintained at a sufficiently high temperature (minimum is 80 °C) for the PDMS to quickly solidify.
13. The heating step should be exactly 30 min. A shorter time may lead to the collapse of the channel, and a longer time would not allow the fluidic layer to successfully bind to the control layer.
14. Place a PDMS slide-size piece of metal on top of the setup to provide additional pressure.
15. If there are leakages in inlets and outlets, seal the bottom of inlets and outlets with PDMS mixture. Keep the chip in the oven for 10 min and test again. If inlets and outlets are blocked, then discard the chip.
16. It's also possible to put the chip into the oven for rapid evaporation. However, you should ensure that the chip has returned to room temperature before surface modification.
17. The modification process is constructed in a sandwich deposition. The first and the second layer (biotin-BSA and neutravidin) are the same for all chip designs. The third layer (antibody) is based on the purpose of experiment. The chip modified by biotin-BSA and neutravidin can be stored in the refrigerator for a maximum of 5 days. However, the antibody needs to be loaded into the channel immediately before the experiment.
18. Three washes are usually enough to remove all unbound cells. However, if you get a large amount of cell pellet, it may be necessary to wash more times. Also, if the volume of the cell pellet is small, you may need to reduce the number of washing step.
19. The 3% BSA in PBS buffer should be stored in the refrigerator. This buffer should be warmed up in the water bath at 37 °C before use.
20. Push hard to remove all bubbles in both the channel and the PTFE tube.
21. This step is prone to the introduction of bubbles. Bubbles may block the channel and then affect the capture. Therefore, we cut the connection part of the buffer syringe and PTFE tube, instead of pulling the syringe out of the tube. Also, in this step, the inlet of PTFE tube needs to remain lower than the chip station at all times.

Acknowledgments

This work was supported by a grant from The CH Foundation.

References

1. Zhang Y, Zhou Y, Li W et al (2018) Multiparameter affinity microchip for early sepsis diagnosis based on CD64 and CD69 expression and cell capture. Anal Chem 90:7204–7211
2. Zhang Y, Lyons V, Pappas D (2018) Fundamentals of affinity cell separations. Electrophoresis 39:732–741
3. Zhou Y, Dong Z, Andarge H et al (2020) Nanoparticle modification of microfluidic cell separation for cancer cell detection and isolation. Analyst 145:257–267
4. Gupta A, Chaudhary VK (2003) Whole-blood agglutination assay for on-site detection of human immunodeficiency virus infection. J Clin Microbiol 41:2814–2821
5. Yüce M, Ullah N, Budak H (2015) Trends in aptamer selection methods and applications. Analyst 140:5379–5399
6. Zhao Y, Xu D, Tan W (2017) Aptamer-functionalized nano/micro-materials for clinical diagnosis: isolation, release and bioanalysis of circulating tumor cells. Integr Biol 9:188–205
7. Song Y, Tian T, Shi Y et al (2017) Enrichment and single-cell analysis of circulating tumor cells. Chem Sci 8:1736–1751
8. Zhang Y, Li W, Zhou Y et al (2018) Detection of sepsis using CD64 expression in a microfluidic cell separation device. Analyst 143:241–249
9. Zhou Y, Zhang Y, Johnson A et al (2019) Detection of culture-negative sepsis in clinical blood samples using a microfluidic assay for combined CD64 and CD69 cell capture. Anal Chim Acta 1062:110–117
10. Zhou Y, Zhang Y, Johnson A et al (2019) Combined CD25, CD64, and CD69 biomarker panel for flow cytometry diagnosis of sepsis. Talanta 191:216–221

Chapter 19

Analgesia and Humane Endpoints for Rodents in Sepsis Research

Christine A. Boehm and Jean A. Nemzek

Abstract

Numerous regulatory bodies around the world require analgesics for rodents undergoing surgery to induce sepsis. Well-controlled pain will decrease morbidity. Options for analgesics include NSAIDs, local analgesics, and opioids. Supportive care can also decrease stress to post-operative animals. As well, humane endpoints should be agreed upon before the study commences so as to alleviate unnecessary pain and distress.

Key words Sepsis, Mouse, Analgesia, Cecal ligation and puncture, Colon ascendens stent peritonitis and fibrin clot implantation, Cecal slurry, Humane endpoints

1 Introduction

In search of better diagnostic and therapeutic tools to combat sepsis, research investigators rely heavily on animal models to replicate the complex immunologic and physiologic responses seen in clinical patients. The vast majority of preclinical studies employ either surgical or nonsurgical rodent models [1]. To optimize their clinical relevance, rodent models often include comorbidities and are used to investigate extreme endpoints including multiple organ failure and survival. Consequently, substantial animal welfare concerns may be raised due to the inherent severity of the disease and the methodologies used to establish these models. Growing evidence that unrelieved pain and distress may impact the scientific results obtained from animal models [2] further heightens the need to address welfare concerns. Therefore, it is imperative that practical recommendations for the relief of pain and distress, including the use of analgesics, adjunct care, and humane endpoints, be available for use in rodent models of sepsis [3–6].

Wendy E. Walker (ed.), *Sepsis: Methods and Protocols*, Methods in Molecular Biology, vol. 2321,
https://doi.org/10.1007/978-1-0716-1488-4_19,

2 Analgesia for Rodents Undergoing Surgery to Induce Sepsis

Analgesia is an attempt to alleviate a painful condition. It is different than anesthesia, which is a controlled and reversible loss of consciousness that makes the patient unable to react to pain. Analgesia is also different than sedation, which makes the animal unaware of its surroundings and drowsy.

The ultimate goal of analgesia is to inhibit the sensation of pain. Well-controlled pain decreases morbidity [2]. Additionally, unalleviated pain can suppress the immune system, introducing yet another experimental variable [7, 8]. However, with sepsis models, analgesia may be seen as a confounding factor, especially if anesthesia and analgesic protocols are not reported in the literature [9, 10].

2.1 Regulatory Concerns

Numerous regulatory agencies have weighed in on the need for analgesics in rodent studies as the standard for veterinary care, putting the burden on the Institutional Animal Care and Use Committee (IACUC) to either require analgesics or scientific justification for the omission of analgesics. In the USA, most IACUCs will categorize experimental rodents according to USDA categories. While the categories were originally intended for use with species other than mice and rats, it is a useful classification that many IACUCs have extended to mice and rats. The USDA pain categories are as follows.

IACUCs will categorize most sepsis surgeries as category D, and sepsis surgeries without analgesics are characterized as category E.

Any institution that receives NIH funds for animal research must have an Assurance on file with OLAW, requiring institutions to adhere to the Public Service Policy on Humane Care and Use of Laboratory Animals. This policy states that "...procedures that may cause more than momentary or slight pain or distress to animals will be performed with appropriate sedation, analgesia, or anesthesia" unless otherwise approved by the IACUC.

As well, the *Guide for the Care and Use of Laboratory Animals*, which is one of the guiding documents for OLAW and AAALAC, Inc., cites the U.S. Government Principles of Utilization and Care of Vertebrate Animals Used in Testing, Research, and Training (Guide p. 12): "... avoidance or minimization of discomfort, distress, and pain when consistent with sound scientific practices, is imperative. Unless the contrary is established, investigators should consider that procedures that cause pain or distress in human beings may cause pain or distress in other animals ... Procedures with animals that may cause more than momentary or slight pain or distress should be performed with appropriate sedation, analgesia, and anesthesia." [11, 12]. It is clear that our regulatory agencies do expect researchers and IACUCs to minimize pain and distress whenever possible.

Similar ethical oversight is provided in other countries, for example, by the Animal Welfare Ethical Review Body (AWERB) in the United Kingdom. It is imperative that the investigator complies with the ethical regulations and guidelines appropriate for their location.

- Category B: Used for breeding colonies and holding protocols.
- Category C: Experimental manipulations do not cause more than momentary pain or distress.
- Category D: Experimentally induced pain or distress is relieved with analgesics, anesthetics, and other methods.
- Category E: Experimentally induced pain or distress is intentionally not relieved. This requires a scientific justification for unrelieved pain or distress.

2.2 Current Recommendations

Although the use of analgesics in sepsis models has been controversial, over the past decade the views of investigators have begun to align with regulatory agencies. It is now generally considered unacceptable to withhold analgesics in the majority of rodent sepsis models [13]. Most recently, an international panel of experts convened to discuss best practices with regard to sepsis models. The panel concluded that analgesics should be considered for all surgical models of sepsis [1], which would include rodent models of cecal ligation and puncture, colon ascendens stent peritonitis, and fibrin clot implantation. In addition, the aforementioned international panel recommended that analgesics be considered for nonsurgical models (endotoxin, cecal slurry, and inoculation) on a case by case basis [1]. The duration of coverage may vary with the model and institutional requirements, but often is recommended for at least 2 days for surgical models, with additional analgesic given as needed. However, consideration should be given not only to the surgical component of the model but to the pain associated with the septic insult itself. Conditions such as peritonitis or local abscesses may continue to be painful after the early postoperative period and additional analgesia should be considered.

There may be rare instances in which analgesics may interfere with the goals of a scientific study. In those cases, the local IACUC would require strong scientific justification and may ask for a pilot study to determine the extent of the interference. Alternatively, a well-designed study with control groups also treated with analgesics may be appropriate and, depending upon the goals of the study, may make the model more translatable to the human clinical condition. When withholding a specific analgesic is contemplated, all alternatives should be explored including other classes of drugs, routes of administration, and application of adjunct methods to decrease stress.

Pain relief is most effective when initiated prior to the painful stimulus to prevent tissue sensitization, decrease the amount of anesthesia required, and decrease the dose of subsequent analgesics. Therefore, institutional guidelines generally require preemptive administration of analgesic agents.

2.3 Selection of Analgesic Agents

Analgesics fall into a number of different categories, including nonsteroidal anti-inflammatory drugs (NSAIDS), local analgesics, and opioids. Ideally, an analgesic regimen would be multimodal, achieved by treating with more than one class of analgesic agent with different mechanisms.

2.3.1 NSAIDs

This category of analgesics is not commonly used in sepsis research models. Due to their anti-inflammatory effects on prostaglandin synthesis [14], NSAIDS significantly suppress immune responses of interest in sepsis models. In addition, NSAIDS have the potential for gastrointestinal toxicity, altered platelet function, and renal toxicity [15, 16]. These effects may be compounded by dehydration and debilitation in septic animals. Therefore, NSAIDS are rarely recommended for analgesia in sepsis models, although they have been examined in preclinical studies as treatments for sepsis.

2.3.2 Local Analgesia

As part of multimodal anesthesia, a few drops of a local analgesic such as a 50:50 mixture of lidocaine and bupivacaine can be painted or dribbled on the skin incision after it is sutured to help provide relatively immediate incisional pain control. Lidocaine works very quickly, within 5 min, but is short acting (30 min to 1 h), while bupivacaine will take up to 30 min to exert its antinociceptive effect, however it may last up to 12 h. In rare instances in which omission of systemic analgesia is scientifically justified, local analgesia may be considered to provide some relief with minimal to no systemic side-effects [17]. Continuous infusion of local analgesics via osmotic pump has also been described; however, these infusions demonstrated significant suppression of the inflammatory responses which would confound sepsis modeling [18].

2.3.3 Opioids

The most important piece of multimodal analgesia in sepsis models is the addition of opioid analgesics. In general, opioids will provide alleviation of post-operative pain when administered at the correct dose and correct timing.

Historically, the use of opioids in sepsis models was limited due to concerns over immunosuppressive effects [19]. These concerns were largely based on the use of morphine which is no longer recommended for sepsis studies. As shown by Eisenstein et al., morphine was shown to potentiate the septic state, showing enteric bacteria in the liver, spleen, and peritoneum, as well as heightened sensitivity to lipopolysaccharide [20, 21]. While morphine's immunosuppressive effects in murine models is well-documented, there are opioids known to be less immunosuppressive and those most thoroughly examined in surgical sepsis models include buprenorphine hydrochloride, buprenorphine sustained release [22, 23], and tramadol. Tramadol has shown little effect on inflammatory markers in experimental sepsis in rats and mice [24, 25]; however, at doses high enough to provide analgesic efficacy, tramadol increased the toxic side effects such as seizures and tremors in both male and female C57BL/6 mice after laparotomy [26] and increased mortality in a rat CLP model [24]. Buprenorphine preparations are the most commonly used analgesic in rodent sepsis models [1]. They have shown minimal effects on inflammation and survival in several studies [25, 27, 28], although rare pilot studies have shown confounding effects on models [17]. A table of frequently used opioids, dosages and routes of administration has been provided for mice (Table 1) and for rats (Table 2). Butorphanol should be used with care in a sepsis models, as it has only been evaluated in tail flick and hot plate assays, and not sepsis models [29]. While several other opioids may be available, they have not been listed due to lack of testing for pharmacokinetic clearance or analgesic efficacy, extremely rapid clearance, or commercial unavailability. Selection and dosing recommendations may vary with sex, strain, and institutional guidelines [20]. As always, when choosing analgesics, it is best to consult with your facility veterinarian. Analgesic regimens for mice and rats are presented in Tables 1 and 2, respectively.

Table 1
Analgesic regimens for mice

Drug	Dose range	Route	Frequency
Buprenorphine, sustained release	1.2–2.0 mg/kg	IP, SC	48–72 h
Buprenorphine hydrochloride	0.1–2.0 mg/kg[a]	IP, SC	4–6 h
Butorphanol	5.0 mg/kg	IP, SC	1–2 h

[a]Note that male C57BL/6 have an increased mortality at the 0.1 mg/kg dose; a pilot study starting with a dose of 0.05 mg/kg for male mice may be warranted

Table 2
Analgesic regimens for rats

Drug	Dose range	Route	Frequency
Buprenorphine, sustained release	1.2 mg/kg	IP, SC	48–72 h
Buprenorphine hydrochloride	0.2–0.5 mg/kg	IP, SC	6–8 h
Butorphanol	2.0 mg/kg	IP, SC	1–2 h

2.4 Assessing Pain Relief

In sepsis models, the assessment of analgesic efficacy may be difficult due to the similarities in signs of pain and the sickness behavior associated with infection. Therefore, general appearance and spontaneous behavioral patterns may not help distinguish the effectiveness of analgesics at the cageside. Nociceptive testing and measurement of physiologic or biochemical markers may not be practical for rapid evaluation of individual mice during sepsis studies. Therefore, the welfare goals for sepsis studies should include using the correct dosage of an analgesic with tested efficacy coupled with adjunct treatments to decrease stress and application of appropriate humane endpoints.

2.5 Adjunct Treatments

Other methods to decrease stress on animals in sepsis studies include the following.

- Limiting the amount of handling that is needed for repeat injections.
- Providing easily accessible food or a diet gel.
- Providing sufficient or increased nesting for thermoregulation.
- Housing in social groups, as rodents are considered to be social animals.

Researchers working on these delicate models should consult their institutional veterinarian to discuss the options that are available at their facility to increase the welfare of animals used in sepsis studies.

3 Humane Endpoints for Animals in Sepsis Experiments

Humane endpoints for sepsis studies can vary greatly. However, criteria for when the animals are euthanized must be clearly defined. The criteria that are chosen for an individual study will vary depending on the type of model, the data that will be collected, and the institution's IACUC guidelines. Although survival studies are a mainstay in sepsis research, the use of "death as an endpoint" has fallen out of favor as ethically unacceptable and should be prevented. Ideally, humane endpoints will preempt animal pain and suffering while still providing scientifically sound endpoints for data collection.

Humane endpoints may be physical, biochemical, behavioral, or a combination of those factors. Ideally, endpoints are study-specific and a pilot study may be required to define the most appropriate endpoints. The methods for assigning useful endpoints through pilot studies have been outlined [5] and investigators are encouraged to work with their facility veterinarian to define appropriate endpoints. The endpoint criteria and analgesic regimen should be reported along with other methods in scientific manuscripts, as described in the Animal Research: Reporting of in Vivo Experiments (ARRIVE) guidelines [30].

Many institutions are moving to score sheets such as those below to assign objective criteria to what was once a subjective process. While the models below help standardize euthanasia criteria in studies, these tools can be further customized as needed to meet each institution's needs.

3.1 Nonsurgical Models

Nonsurgical models of sepsis do not have the complicating factor of surgical recovery as a variable, however sepsis itself still requires frequent monitoring of objective criteria to ensure animals are euthanized at an appropriate time-point. In a model of septic pneumonia, the M-CASS scoring system was developed and tested. It assigned a score from 1 to 4 for each parameter assessed (fur aspect, activity, posture, behavior, chest movements, chest sounds, eye lids, body weight loss score, and monitoring frequency) [31]. When mice reached a score of 4 for all eight parameters, they were euthanized. However, if a mouse was not active when stimulated, it was collapsed in a head down position, or was gasping and had its eyes closed when stimulated, it was also euthanized. As well, mice that had reached a score of 4 for two of the items at the last monitoring time point of the day were also euthanized. Mice were evaluated every 12 h in this study.

Another popular scoring system for humane endpoints in a nonsurgical CLP model is the Murine Sepsis Score developed by Shrum et al. This tool measured seven parameters (appearance, level of consciousness, activity, response to stimulus, eyes, respiration rate, and respiration quality) [32] every 2 h after the induction of sepsis, and hourly afterward for 24 h. Mice were euthanized if the MSS reached 21 or greater, or if the respiratory parameters reached a score of 3.

3.2 Surgical Models

Similar endpoints can be used for surgical sepsis models. Studies by Nemzek et al. have found that a decreased body temperature, weight gain and elevated plasma IL-6 levels serve as early indicators of impending death after cecal ligation and puncture [5]. However, these may not pinpoint the timing of death with sufficient accuracy for some studies. The inability to ambulate and loss of righting reflex may provide greater accuracy in prediction of death but occur late, in the moribund state, and therefore, provide less relief. Ideally, endpoints should be selected to provide the earliest relief possible and this may require pilot studies in order to support sound scientific endpoints.

3.3 Neonatal Considerations

Neonatal models may require different monitoring parameters. As reported by Brook et al., it is recommended to monitor the mice at 2 h post-challenge, then every 4–6 h for the first 2 days [33]. An exception can be made for overnight monitoring, when the monitoring interval can be extended up to 8 h. Many lab and IACUCs may extend the 8 h monitoring to a 12 h period to reflect a more practical schedule. To monitor the mice, it is recommended to go through a series of tests. The first test involves measuring the neonate's righting reflex when laid on both the right and left side as well as mobility, and score the results. When the neonate has reached a humane endpoint, it is euthanized. It is important to return viable neonates to the same dam and reform the nest to prevent scatter of the neonates.

4 Summary

In summary, the appropriate use of analgesics and humane endpoints benefits the health and welfare of rodents used in sepsis research. While the choice of analgesic drugs may be study specific, preemptive buprenorphine appears to cause minimal effects in many rodent models of sepsis. Clearly defined humane endpoints may be easily applied to sepsis studies in systematic scoring systems and are a necessity particularly in survival studies. In all cases, investigators are encouraged to discuss measures to relieve pain and distress with their institutional veterinarian and to document refinements to insure reproducibility of studies. Ultimately, these measures should reduce the physiological artifacts induced by unmitigated pain and distress, resulting in improved animal models and higher quality scientific studies.

References

1. Zingarelli B, Coopersmith CM, Drechsler S et al (2019) Part I: minimum quality threshold in preclinical sepsis studies (MQTiPSS) for study design and humane modeling endpoints. Shock 51:10–22
2. Peterson NC, Nunamaker EA, Turner PV (2017) To treat or not to treat: the effects of pain on experimental parameters. Comp Med 67:469–482
3. Carpenter KC, Hakenjos JM, Fry CD et al (2019) The influence of pain and analgesia in rodent models of sepsis. Comp Med 69:546–554
4. Nemzek JA, Hugunin KM, Opp MR (2008) Modeling sepsis in the laboratory: merging sound science with animal well-being. Comp Med 58:120–128
5. Nemzek JA, Xiao HY, Minard AE et al (2004) Humane endpoints in shock research. Shock 21:17–25
6. Jeger V, Hauffe T, Nicholls-Vuille F et al (2016) Analgesia in clinically relevant rodent models of sepsis. Lab Anim 50:418–426
7. Page GG (2005) Surgery-induced immunosuppression and postoperative pain management. AACN Clin Issues 16:302–309; quiz 416-8
8. Page GG, Blakely WP, Ben-Eliyahu S (2001) Evidence that postoperative pain is a mediator of the tumor-promoting effects of surgery in rats. Pain 90:191–199
9. Anderson SL, Duke-Novakovski T, Singh B (2014) The immune response to anesthesia: part 1. Vet Anaesth Analg 41:113–126
10. Anderson SL, Duke-Novakovski T, Singh B (2014) The immune response to anesthesia: part 1. Vet Anaesth Analg 41:553–566
11. Albus U (2012) Guide for the care and use of laboratory animals (8th edn), (ed). SAGE Publications Sage UK, London, England
12. IRAC (1985) US government principles for utilization and Care of Vertebrate Animals Used in testing, research, and training. Prepared by the Interagency Research Animal Committee, National Institutes of Health

13. Stortz JA, Raymond SL, Mira JC et al (2017) Murine models of sepsis and trauma: can we bridge the gap? ILAR J 58:90–105
14. Narver HL (2015) Nalbuphine, a non-controlled opioid analgesic, and its potential use in research mice. Lab Anim 44:106–110
15. Clark TP (2006) The clinical pharmacology of cyclooxygenase-2–selective and dual inhibitors. Veterinary clinics: small animal. Practice 36:1061–1085
16. Fish R, Danneman PJ, Brown M et al (2011) Anesthesia and analgesia in laboratory animals. Academic Press, Cambridge
17. Chen W, Brenner M, Aziz M et al (2019) Buprenorphine markedly elevates a panel of surrogate markers in a murine model of sepsis. Shock 52:550–553
18. Gallos G, Jones DR, Nasr SH et al (2004) Local anesthetics reduce mortality and protect against renal and hepatic dysfunction in murine septic peritonitis. Anesthesiology 101:902–911
19. Rittner HL, Roewer N, Brack A (2010) The clinical (ir)relevance of opioid-induced immune suppression. Curr Opin Anaesthesiol 23:588–592
20. Eisenstein TK, Meissler J, Rogers TJ et al (1995) Mouse strain differences in immunosuppression by opioids in vitro. J Pharmacol Exp Ther 275:1484–1489
21. Hilburger ME, Adler MW, Truant AL et al (1997) Morphine induces sepsis in mice. J Infect Dis 176:183–188
22. Foley PL, Liang H, Crichlow AR (2011) Evaluation of a sustained-release formulation of buprenorphine for analgesia in rats. J Am Assoc Lab Anim Sci 50:198–204
23. Kendall LV, Hansen RJ, Dorsey K et al (2014) Pharmacokinetics of sustained-release analgesics in mice. J Am Assoc Lab Anim Sci 53:478–484
24. Nardi GM, Bet AC, Sordi R et al (2013) Opioid analgesics in experimental sepsis: effects on physiological, biochemical, and haemodynamic parameters. Fundam Clin Pharmacol 27:347–353
25. Hugunin KM, Fry C, Shuster K et al (2010) Effects of tramadol and buprenorphine on select immunologic factors in a cecal ligation and puncture model. Shock 34:250–260
26. Wolfe AM, Kennedy LH, Na JJ et al (2015) Efficacy of tramadol as a sole analgesic for postoperative pain in male and female mice. J Am Assoc Lab Anim Sci 54:411–419
27. Cotroneo TM, Hugunin KM, Shuster KA et al (2012) Effects of buprenorphine on a cecal ligation and puncture model in C57BL/6 mice. J Am Assoc Lab Anim Sci 51:357–365
28. Herndon NL, Bandyopadhyay S, Hod EA et al (2016) Sustained-release buprenorphine improves postsurgical clinical condition but does not alter survival or cytokine levels in a murine model of polymicrobial sepsis. Comp Med 66:455–462
29. Gades NM, Danneman PJ, Wixson SK et al (2000) The magnitude and duration of the analgesic effect of morphine, butorphanol, and buprenorphine in rats and mice. Contemp Top Lab Anim Sci 39:8–13
30. Percie du Sert N, Hurst V, Ahluwalia A et al (2020) The ARRIVE guidelines 2.0: updated guidelines for reporting animal research. J Cereb Blood Flow Metab 40:1769–1777
31. Huet O, Ramsey D, Miljavec S et al (2013) Ensuring animal welfare while meeting scientific aims using a murine pneumonia model of septic shock. Shock 39:488–494
32. Shrum B, Anantha RV, Xu SX et al (2014) A robust scoring system to evaluate sepsis severity in an animal model. BMC Res Notes 7:233
33. Wilcox ME, Rubenfeld GD, Walczak KD et al (2019) Actigraphic measures of sleep on the wards after ICU discharge. J Crit Care 54:163–169

Chapter 20

Agent-Based Modeling of Systemic Inflammation: A Pathway Toward Controlling Sepsis

Gary An and R. Chase Cockrell

Abstract

Despite nearly 50 years of research there currently remains no mediator-directed therapy approved for the treatment of sepsis. The failure to effectively translate the copious mechanistic knowledge regarding systemic inflammation to effective therapies is a dramatic example of the translational dilemma. Dynamic computational modeling has been proposed as a vital means of integrating community-wide knowledge into an investigatory framework that allows the application of engineering-like principles to the problem of sepsis. Agent-based modeling is a computational modeling method that has been used to address some of the fundamental issues facing the sepsis research community. This chapter will introduce the rationale to augment traditional research practices with agent-based modeling, describe the basic steps in the construction and use of agent-based models, and provide examples of how the use of agent-based modeling can provide an investigatory pathway to solving the challenge of sepsis.

Key words Agent based modeling, Sepsis, Multiple organ failure, Acute inflammation, Systemic inflammation, Mathematical models, Multiscale models, Translational systems biology, Translational research, Knowledge representation, Machine learning, Artificial intelligence, In silico trials

1 Introduction

1.1 The Unmet Challenge of Sepsis

Sepsis refers to a series of acute inflammatory processes and their pathophysiologic consequences that occur in response to severe infection. Sepsis is responsible for more than 215,000 deaths in the USA per year with an annual healthcare cost of over $16 billion [1]. While care process improvements in the treatment of sepsis, such as the development of treatment bundles and practice guidelines, have improved clinical outcomes in the past few decades, the attempt to find new drugs to treat the underlying biological abnormalities driving sepsis has been marked by utter failure: there is currently not a single drug approved by the US Food and Drug Administration that targets the underlying pathophysiology of sepsis [2, 3]. Though the basic mechanisms and core mediators/cytokines involved in acute inflammation have been known for

Wendy E. Walker (ed.), *Sepsis: Methods and Protocols*, Methods in Molecular Biology, vol. 2321,
https://doi.org/10.1007/978-1-0716-1488-4_20,

decades, controlling the complex, nonintuitive, and paradoxical nature of the acute inflammatory response in critical illness has defied the reductionist approaches currently employed by the general biomedical research community. Sepsis researchers have produced numerous potential interventions over the last several decades that were promising at the preclinical level: these included manipulation of various cytokines/mediators, damage-associated molecular pattern molecules, oxygen and nitrogen free radicals, coagulation pathway intermediates, and vasoactive peptides and lipids, nearly all of them predicated on the initial assumption that the treatment for sepsis involved somehow attenuating the pro-inflammatory response. Unfortunately, none of these interventions have been deemed successful in Phase III clinical trials. These failures led some to question whether anti-inflammatory biologics were properly applied to sepsis [4, 5]; however these critiques did not fundamentally address the essential issue at hand: namely how to actually translate preclinical knowledge efficiently (and successfully) into effective clinical therapeutics.

We have termed this gap between basic mechanistic understanding and the effective development of clinical therapeutics as the translational dilemma [6], and proposed an overall research strategy, translational systems biology [7, 8], that aims to overcome the current limitations of the biomedical research process. Translational systems biology is an engineering-inspired approach [8] centered on dynamic, mechanistic computational models with direct clinical interpretations and applications. The use of translational systems biology to study sepsis involves using computational models as a means of dynamic knowledge representation to provide additional information about the systems-level behavior of acute inflammation beyond what could be gleaned from preclinical animal models. This included the execution of in silico clinical trials of existing and putative mediator-based interventions for sepsis, with the demonstration that these models could predict failure when preclinical studies showed promise [9, 10], and, more recently, the application of machine learning, evolutionary computing and model-based artificial intelligence to actually define what is needed to control sepsis [11, 12]. This chapter will describe the use of one of the central methods of translational systems biology, agent-based modeling, and how it can be used to augment traditional modalities of research into sepsis.

1.2 Dynamic Knowledge Representation Via Agent-Based Modeling (See Note 1)

The Translational Dilemma can be thought of as a paradoxical effect of the success of biomedical research: as greater amounts of information become available describing increasingly finer levels of biological mechanism, it is increasingly difficult for researchers to effectively integrate that information into a clinically relevant system-level context [6]. Researchers may construct intricately detailed diagrams that can fill entire office walls, but when it comes time to assess how such a complex construct actually works one necessarily utilizes intuition and an ill-defined "sense" of how all those arrows and boxes fit together. The insufficiency of these informal and subjective conceptual models is immediately evident when they break down through the necessary process of hypothesis falsification intrinsic to science: without a formal representation of what is actually being tested, successful falsification has no rigorous means of determining the next investigatory step. Translational systems biology aims to overcome this limitation by developing methods of dynamically representing knowledge to allow researchers to express and evaluate conceptual models more effectively. Computer modeling can be seen as a means of dynamic knowledge representation by providing a formal means of testing, evaluating and comparing what is currently known within the research community. In order to be able to "see" the consequences of a particular hypothesis within a broader structure/conceptual model, the formally represented knowledge is moved from a static depiction of relationships as depicted in a flowchart or state diagram to a dynamic model wherein the mechanistic consequences of each hypothesis can be observed and evaluated.

Agent-based modeling (ABM) is an object-oriented, discrete event and spatially explicit computational modeling technique that is centered on representing the behaviors and interactions of the individual components of a system [13]. It has characteristics that make it well suited for creating aggregated modular multiscale models [14, 15]; namely, ABM focuses on the rules and interactions between the individual components of a system, generating populations of those components and simulating their interactions in a "virtual world" to create an in silico experimental model [13, 16–19]. ABM rules are often expressed as conditional statements ("if-then" statements), which make ABM suited to translating the hypotheses (expressed in natural language) that are generated from basic science research. There are four characteristics of ABM that deserve particular emphasis.

1. ABMs are *spatial*. Agent-based modeling has its origins in 2-dimensional cellular automata, and as such many ABMs are "grid-based." The inherently spatial nature of ABMs makes them well suited to representing structural relationships in a system under study. Nonmathematicians can model fairly complex architectures and topologies with greater ease and flexibility than may be possible with partial differential equations, leading to more intuitive knowledge translation into a model. Furthermore, the spatial nature of ABMs facilitates implementing the concept of *local knowledge*, where individual agents are only able to sense and respond to cues in their local environment (as is generally the case in biological systems; even neural transmissions involve local interactions between the ends of

neurons). This locality allows for internal heterogeneity of system behavior, which produces population level heterogeneity as well.

2. ABMs utilize *parallelism*. This property of ABM sets it apart from other object-oriented modeling methods such as Petri nets or finite state machine models. In an ABM each agent class has multiple instances within the model, forming a population of agents that interact in an emulated (usually) parallel processing environment. Within the execution of an ABM, heterogeneous individual agent behavior within a population of agents results in systemic dynamics that result in observable output that mirrors the behavior at the higher-hierarchical level. A classic example of this is how relatively simple interaction rules among birds can lead to sophisticated flocking patterns. This concept is readily translated to biology, where there are recognized classes/types of cells, which share some common level of functionality, but the individual cells themselves may exhibit a variety of behaviors based on their local stimuli.
3. ABMs utilize *stochasticity.* Many systems, particularly biological ones, include behaviors that appear to incorporate randomness. "Appear to" is an important distinction, since what may appear to be random is actually deterministic from a mathematical standpoint. However, from a practical point of view, despite the fact that a particular system may follow the rules of deterministic chaos, at a higher-order observational level it is impossible to actually define the initial conditions from whence its behavior evolves. ABMs address this issue via the generation of populations of agents. Once one is dealing with populations, then it is possible to establish probabilities of a particular behavior for the population as a whole, and therefore also a probability function for the behavior of a single agent. This probability function is incorporated into the agent's rules. When instantiated and run in parallel with other agents, each agent follows a particular trajectory of behavior as its behavior rules' probabilities "collapse" with each step of the model's run. This stochasticity allows the generation of a "population" of behavioral outputs from a single ABM, reflecting the heterogeneity present in a biological population and allowing one to move beyond the "behavior curves" seen in differential equation models toward "behavior spaces" more consistent with biological observation.
4. ABMs reproduce *emergent properties.* Due to the parallelism, intrinsic stochasticity, and enforcement of locality resulting from its spatial architecture, a central hallmark of ABMs is the fact that they generate systemic dynamics that often could not have been reasonably inferred from examination of the rules of the agents, resulting in the so-called emergent behavior. To

return to the example of the bird flock, superficial observation would seem to suggest the need for an overall "leader" to generate flock behavior, requiring rules for flock-wide command and control communication. This, however, is not true; birds function on a series of locally constrained interaction rules and the flocking behavior emerges from the aggregate of these interactions. The capacity to generate emergent behavior, often paradoxical and nonintuitive, is a vital advantage of using an ABM to simulate complex systems.

ABMs have been used to study biomedical processes such as sepsis [13, 18], cancer [15, 20], inflammatory cell trafficking [21, 22], wound healing [23], and intracellular structure and signaling [24–26]. In general, most biomedical ABMs focus on cells as the primary agent level (with notable exceptions [24–26]). Cells are a natural agent level dictated by the organizational structure of biology, and from a knowledge translation standpoint, form a ready level of "encapsulated complexity" that can be addressed with relatively straightforward input–output rules. Furthermore, while the number of cells present in an organism is considerable, it is still magnitudes less than the number of molecules involved in intracellular signaling. Due to their spatial and structural relationships, cellular populations are less amenable to the application of the mean field approximations and mass action kinetics that provide the basis for effective ordinary differential equation models. This chapter will describe the basic steps of how to develop an ABM, using sepsis as an example, and finish with a discussion of the role of dynamic computational modeling in addressing the challenge of sepsis. For a summary of the strengths and limitations of agent based modeling (*see* **Note 1**).

1.3 The Innate Immune Response Agent-Based Model of Sepsis

The Innate Immune Response agent-based model (IIRABM) is one of the original ABMs used in the biomedical arena [9]. Its development was begun in the late 1990s [18] as a means of integrating then state-of-the-art knowledge about the cells and cytokines involved in the acute inflammatory response with the specific goal of creating surrogate in silico clinical populations. It was originally developed to demonstrate the use of in silico clinical trials as a means of evaluating the plausibility of planned potential interventions [9], but has since proved valuable in providing fundamental insights into the mathematical and dynamic properties of sepsis that account for patient heterogeneity [27], demonstrating the futility of standard biomarker-based outcome prediction [27], and served as a proxy model [28] for control discovery for sepsis. This most recent control discovery work has employed advanced computational methods such as genetic algorithms/evolutionary computing [11] and deep reinforcement learning/artificial intelligence [12] in order to define the boundary conditions of what it would actually take to treat sepsis. Even though the IIRABM is nearly 20 years old its core component structure remains valid; perhaps even more valid than when it was originally developed as it has predicted a series of behaviors associated with sepsis that have since been recognized in the subsequent years, specifically the temporal concurrence of pro- and anti-inflammatory cytokine responses (as opposed to sequential pro- and compensatory responses) [29, 30] and the importance of the immunoparalyzed recovery phase of sepsis, particularly with respect to its prolonged duration [31–34]. The robustness of this model demonstrates the power of abstraction and dynamic modeling of complex biological phenomena such as sepsis, and of the benefit of Translational Systems Biology in the study of sepsis. The following sections will provide a more detailed description of the IIRABM and present the modeling steps used in its construction.

2 Methods

2.1 Steps in the Development and Use of an Agent-Based Model

Many of the basic principles for developing a biomedical computational model are applicable to the construction and use of an ABM. These steps typically involve: (1) delineation of the system being modeled, (2) determination of the intended use of the model and (3) the suitability of the modeling method in question to the answers to #1 and 2. As mentioned above, the structure of an ABM facilitates its translational use for modeling mechanistic information expressed in natural language, and as such it is often more intuitive for nonmathematicians to grasp. The following sections will use the IIRABM as an example of how to develop an ABM. For a summary of general modeling principles and potential pitfalls applicable to the following steps (*see* **Note 2**).

2.2 Defining the Purpose of the Model

This is a very basic, but often overlooked aspect of model construction. The question serves to remind us that models should not be created just because they can be; the justification of their development must be framed as to serve some particular purpose, even if that purpose is merely to demonstrate the capability of a particular type of model construction. Answering this question explicitly sets the groundwork for expectations with respect to interpretations of the model's output, and any conclusions that can be drawn from its behavior.

The IIRABM model was developed to serve two purposes. First, it was initially intended as a "proof-of-concept" model, to demonstrate that agent-based modeling was a viable means of dynamically representing mechanistic biological knowledge: specifically that aggregated cellular interactions could be used to simulate virtual patients. Second, the intent of the IIRABM was to demonstrate the potential of in silico clinical trials as a means of testing the plausibility of any planned intervention, by simulating a series of existing and planned anticytokine clinical trials. This last goal led to several modeling decisions:

1. Since the goal of the IIRABM was to integrate cellular information in as similar a fashion as in the real world (where interacting cells generate organismal behavior), this meant that some set of fundamental properties of the overall real world system (the organism) needed to be accounted for, distinct from the individual rules for the cellular agents. In more technical terms, it is necessary to identify a system-level phenotype or *macrostate* metrics that distinguish the model from its individual components (termed *microstate*). We will discuss this in more detail in the next section.
2. Since the intent of the IIRABM was to simulate planned cellular and molecular interventions, the IIRABM needed the representational capacity to reflect those interventions. For instance, anti–tumor necrosis factor interventions were one of the first anticytokine trials to be simulated. Therefore, the dynamics and effect of Tumor Necrosis Factor needed to be incorporated into the IIRABM.
3. All models represent some abstraction of the real world. It therefore follows that an appropriate level of abstraction should be selected for the IIRABM. While it is difficult (if not impossible) to define specific guidelines for how to determine the appropriate level of abstraction, I would pose two general principles one should keep in mind: First, the classical modeling adage "KISS = Keep It Simple Stupid," which is a variant of Occam's razor: namely that one should aspire to the simplest model able to serve the purposes of the modeling task. This is crucial to remember, as it helps avoid the natural tendency of biological modelers to "put it all in," that is, the desire to add every conceivable detail with the presupposition that every bit is important. Second, avoid abstractions that embed assumptions of effect. This is a counterweight to KISS, and an important one when dealing with complex systems: it is exactly the inability to intuit an effect that defines this class of systems. Examples of this include assuming that a phenomenological effect is a mechanistic one. As is the case when dealing with information from knock-out experiments, very often there is an assumption of linear causality by the deleted gene, where in fact the knockout organism represents a system-level compensation for the absence of that particular gene. These two principles represent a push-pull when picking a level of abstraction for any model, though cell-as-agent-level ABMs present a natural means of accommodating this decision, as cells form a natural level of encapsulated complexity in biological systems. The

choice of cell-as-agent-level essentially allows the modeler to treat cells as black-box, input–output objects, which maps fairly readily to how much basic biomedical research is presented (*see* **Note 2**).

2.3 Defining the Reference and Conceptual Models (See Note 3)

The reference model is a real world system that defines the informational basis of the ABM. The conceptual model is the necessary abstraction of that reference model that allows it to be represented and evaluated: its topology, the agents and a starting point for identifying the literature basis of the agent rules. Note that the reference model may be a particular experimental preparation or an aggregated conceptual model in the mind of the researcher. If the latter is the case, then it is important to define as explicitly as possible the knowledge foundation of the conceptual model, particularly with respect to capabilities and limitations of the wet lab experiments that provide the basis of the conceptual model.

As an example, the reference model for the IIRABM is systemic inflammation. The conceptual model views disordered systemic inflammation as being propagated by interactions between the endothelium and circulating inflammatory cells. This was one of the prevalent views concerning the pathophysiology of sepsis in the late 1990s and endothelial dysfunction remains a component of modern concepts of sepsis. Notably, the IIRABM does not explicitly deal with tissue/organ damage, and therefore is not able to simulate organ dysfunction, failure or support (though it was integrated with an epithelial barrier model to bring in this level of representation in [35]). As such, the IIRABM is unable to produce/replicate physiological output, such as vital signs, and interpretation of the output of the IIRABM necessarily requires a more abstract representation of what a "damaged" system consists of. Therefore, because it is essentially an endothelium-based representation, the macroscopic system-level "health" of the IIRABM is reflected as the aggregate amount of endothelial dysfunction/damage as reflected by a perfusion variable called "oxy." The simulated system dies when total damage (defined as aggregate endothelial cell damage) exceeds 80%; this threshold represents the ability of current medical technologies to keep patients alive (i.e., through organ support machines) in conditions that previously would have been lethal.

Even more critically, it is important to be cognizant of what the overall system is intended to represent: an abstraction of an organism that has a high degree of baseline dynamic stability, for example "health." It is an essential goal of utilizing biomedical ABMs that they must include a representation of baseline health, and that disease states occur as a result of perturbations which disrupt that health. To do otherwise is to create a disease model that has no grounding of the healthy condition that a potential therapy is intended to restore. This requirement that the ABM must be dynamically stable at a baseline-unperturbed state AND be able to tolerate some level of perturbation without progressing to total system collapse places a considerable constraint on the construction of the ABM, including what rules/pathways/functions need to be included in the model (*see* Subheading 2.6).

2.4 What Is the Topology of the Model?

As mentioned above, many ABMs are based on two-dimensional grids in which the edges "wrap" to form tori. Other potential topologies include three-dimensional, cube-based structures and various network structures (such as scale-free, giant component or small world configurations). In general, two-dimensional grids are sufficient to represent systems in which there is primarily one plane of agent interactions (along the surface of the grid), though it is possible to model multiple "layers" of data at a particular grid-square (akin to the data structures of Geographical Information Systems (GIS)) (Fig. 1 depicts an example of this from an ABM of host-pathogen interactions associated with gut-derived sepsis [36]).

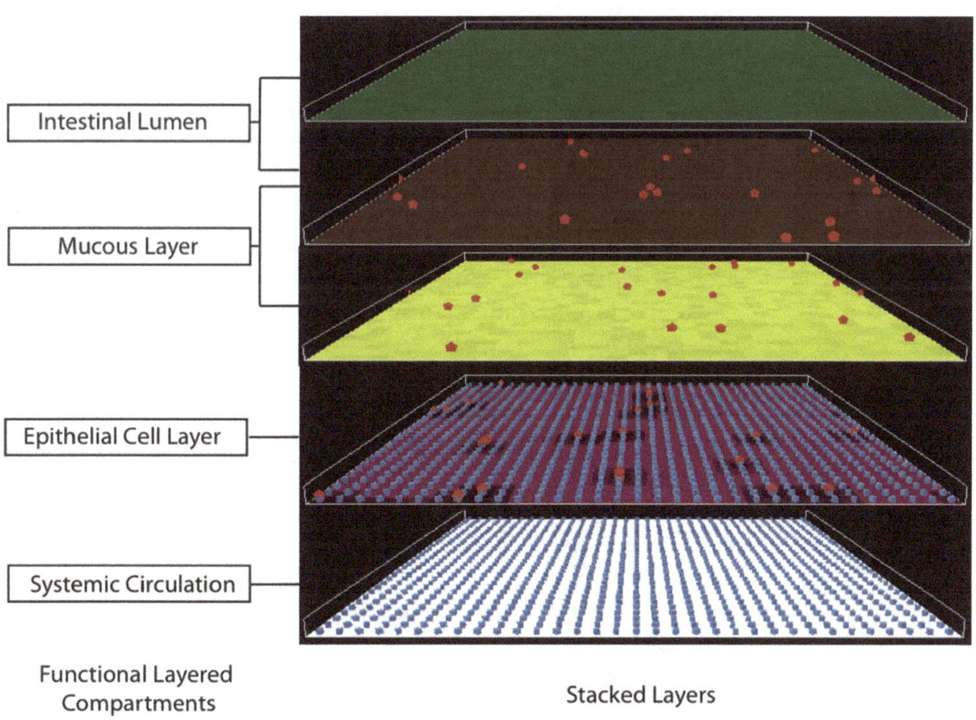

Fig. 1 Multi-data layer Structure of a two-dimensional ABM: This figure from an ABM that models the multiple interacting layers associated with gut-derived sepsis (from Ref. 36, NOT the IIRABM), depicts an example of how multiple data layers can be projected onto a shared two-dimensional plane, akin to how modern map applications can layer multiple data structures (i.e., road networks, satellite images, geographical elevations, etc.). (Figure reproduced from Ref. 36 under Open Access Creative Commons License)

As an example, the IIRABM is a two-dimensional abstract representation of the human endothelial-blood interface. This abstraction is designed to model the endothelial-blood interface as the initiation site for acute inflammation and represents this interface as the unwrapped internal vascular surface of a 2D projection of the aggregation of the sum total of the body's capillary network. The closed circulatory surface can be represented as a torus, and this two-dimensional area makes up the space that is simulated by the IIRABM. The spatial scale of the real-world system is not directly mapped using this scheme. This abstraction serves two primary purposes: to allow circumferential access to the injury/infection by the innate immune system, and to incorporate multiple levels of interaction between leukocytes and tissue.

2.5 Defining the Agents (See Note 2)

This is often the critical question and decision when constructing an ABM. Agents need to be a well-circumscribed group of components that can be treated as input–output objects following similar (if not identical) state-transition rules. The state of an agent is determined by a series of state variables internal to the agent (i.e., receptor binding status, gene activation levels, etc.), which are then modified based on external state variables (i.e., mediators/cytokines) in some spatially defined interaction environment for the agent. Therefore, selection of an agent level necessarily leads to some "compression of complexity" of the internal workings of the agent, an assumption implies that the informational basis of the input–output rules are valid irrespective of the particular mechanisms internal to the agent (the "black box" phenomenon). This is a critical point in determining the agent level. In general, the intended use of the model, vis a vis a planned intervention or particular targeted mechanism for study, will determine the resolution or *granularity* of the ABM. At the granularity chosen there needs to be a fairly certain linear approximation of mechanistic causality: that is, how certain are you that state variable a goes to state variable a' with mechanism b? Explicit delineation of the granularity of the model and the corresponding assumptions are critical in avoiding petitio principii, or "programming the proof." This can manifest either as treating the agent as a "black box" and focusing purely on its response as an input–output object, or, more commonly, with some degree of abstraction in respect to the progression of its internal state variables, such as by abstracting signaling and synthetic pathways.

As an example, the IIRABM is a cell-as-agent-level ABM, incorporating endothelial cells and subsets of circulating inflammatory cells: monocytes/macrophages, neutrophils, TH0, TH1, and TH2 cells as well as their associated precursor cells. Topologically, the endothelial cells form the "surface" of the IIRABM, whereas the inflammatory cells, in their inactivated form, move over that surface until they are signaled to begin adhesion to activated endothelial cells. As noted above, little to none of the internal regulatory machinery of these cells are represented; rather they are primarily modeled as reactive objects that change state, move or produce mediators based on environmental variables/signals (see the next section for more details).

Infectious insults to the IIRABM are initiated using five parameters representing the size and nature of the injury/infection as well as a metric of the host's resilience—initial injury size, microbial invasiveness, microbial toxigenesis, environmental toxicity, and host resilience. In previous work [37], a sweep over these parameters determined the plausible boundaries for conditions that could be considered clinically relevant. These are parameter sets that lead to all possible final outcomes when stochastically replicated—complete healing, death by infection, or death from immune dysregulation/sepsis (*see* Subheading 2.9).

2.6 Defining the Agent Rules (See Note 2)

Once the agent level has been selected, attention is then turned to examination of the literature concerning the potential mechanisms to be modeled, and determining an interaction scheme between those mechanisms. It is often useful to express the latter goal in the form of a state or influence diagram that can be used to guide the actual coding of the ABM. The determination of the agent rules forms the primary translational step in the development of an ABM. A qualitative approach is recommended at the outset in order to maintain a clear mapping between the basis of the reference/conceptual model and the ABM computer code. Too great an attention to specific details with respect to kinetic rate constants (for instance) in the initial translation phase can prove to be overwhelming in terms of the complexity of the model. Therefore, it is useful to classify processes into relatively general groups of magnitude. For instance, with respect to determining the "time" it takes for a particular process it is usually sufficient to classify processes as "very fast (order of seconds)," "fast (order of minutes)," slow (order of hours)," and "very slow (order of days to weeks)." It should be acknowledged that this grouping is highly subjective. However, the overall subjectivity of the divisions is less important than (1) consistency within a particular ABM, (2) an awareness of the assumptions implicit upon the choice of the level of granularity, and (3) making this explicit and transparent when communicating the model.

As an example, the IIRABM does not aspire to comprehensively represent every pathway present in the cell types included (*see* Subheading 2.1). Rather, there is a conscious choice made to incorporate those pathways and mediators of particular interest as potential targets for therapies, as well as the compensatory pathways known to provide negative feedback on the behavior of the systems. Here it important to remember that the overall system should be dynamically stable and robust to some level of perturbation (Subheading 2.1), implying that the model should incorporate negative feedback control or a specified degradation rate for signals/processes. Figure 2 provides a state diagram for the IIRABM, depicting the various receptors, functions and cytokines through which the different cell types interact [11]. A verbal description of the various rules can be found in the Appendix of a prior report [13].

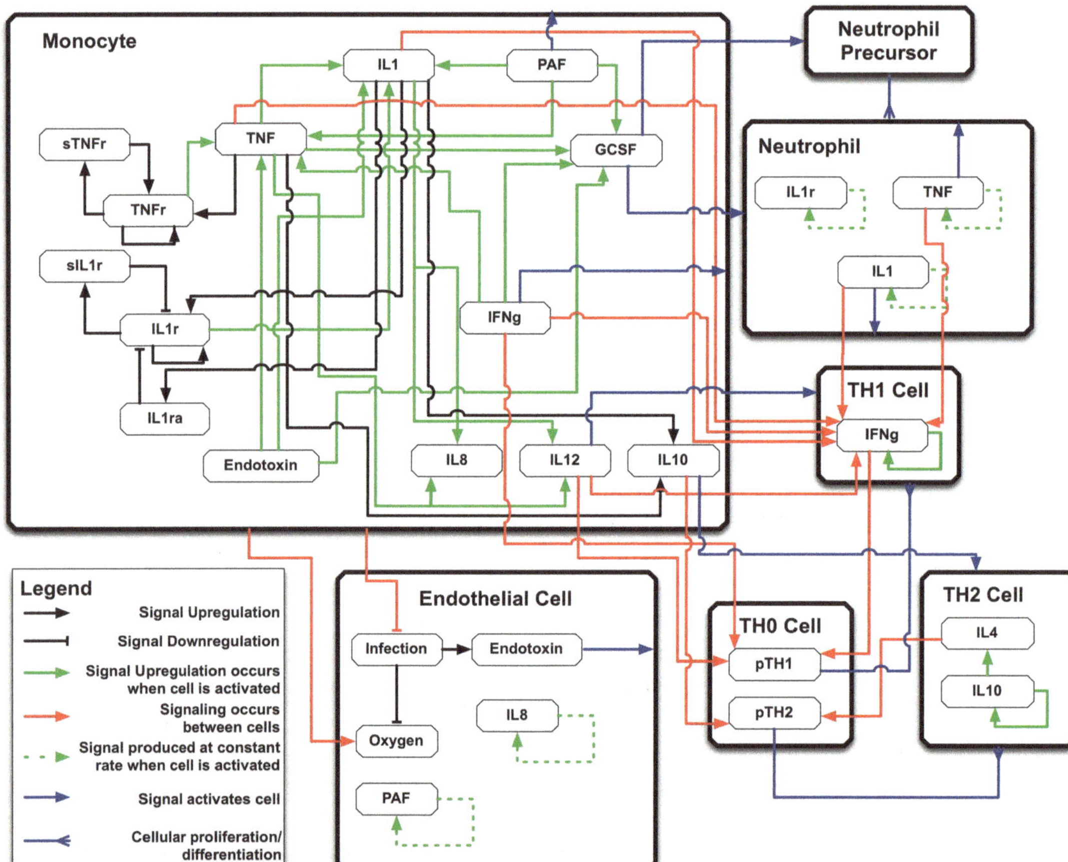

Fig. 2 A state diagram of the cell types and molecular interactions in the IIRABM: This figure is a wire-frame diagram that depicts the primary active cell types (e.g., not bone marrow-derived precursor lineage cells) and some of their primary interactions in the IIRABM. TNF = Tumor Necrosis Factor, TNFr = Tumor Necrosis Factor Receptor, sTNFr = Soluble Tumor Necrosis Factor Receptor, IL1 = Interleukin-1, IL1r = Interleukin-1 Receptor, sIL-1r = Soluble Interleukin-1 Receptor, IL1ra = Interleukin-1 Receptor Antagonist, PAF = Platelet Activating Factor, GCSF = Granulocyte Colony Activating Factor, IFNg = Interferon Gamma, IL4 = Interleukin-4, IL8 = Interleukin-8, IL-10 = Interleukin-10, IL-12 = Interleukin-12, pTH1 = Compilation variable favoring transition from TH0 to TH1 cells, pTH2 = Compilation variable favoring transition from TH0 to TH2 cells. (Figure reproduced from Ref. 11 under Open Access Creative Commons License)

2.7 Putting it Together: Programming the Model (See Note 4)

There are two primary options when it comes time to write the code for an ABM: (1) a stand-alone program can be written in a basic computer language, such as C or C++, or (2) a program can be written using an established ABM modeling toolkit. Option 2 has the advantage that many of the programming underpinnings of ABM, such as object-class definition, emulated parallelization, creation of a graphical user interface and data collection tools, are not trivial programming tasks, and having these issues preaddressed in an established ABM toolkit allows a researcher to focus on the modeling aspect of the project rather than the programming aspect. A list of available ABM toolkits/modeling environments can be seen in **Note 4**. Since ABM is a discrete event modeling system and the program progresses in a stepwise fashion, there must be a selection of the base time interval for each step. Mechanisms to be translated into agent rules are broken into steps based on the duration of those mechanisms, and further translated into program code. It is important to keep in mind that a particular code block will run sequentially (even in an emulated parallel environment), therefore the order or *schedule* of process events needs to free of inadvertent internal feedback loops. For instance, if a particular agent has a rule where it produces an external state variable that in turn affects the agent's subsequent production of that same external state variable, then placing the production code at the beginning of the code block will lead to an artifactual enhancement of any forward feedback effects for that particular rule.

As an example, the IIRABM was constructed using the freeware software toolkit Netlogo [38]. The entire model, along with extensive documentation, is available the Netlogo Community Models Website (http://ccl.northwestern.edu/netlogo/models/community/Innate%20Immune%20Response). The time step for the IIRABM is 7 min per step. The base size of the virtual world is arbitrarily set as a 101 × 101 square 2-dimensional grid. The entire background grid is filled with agents representing endothelial cells. The inflammatory cell species are randomly distributed across the surface and move in a semirandom fashion until they come into contact with any chemotactic factors present in the environment, at which time they follow the steepest gradient of that chemotactic factor. The inflammatory cell populations have a defined life span, and precursor cells present in an abstracted bone marrow compartment. Thus, in an unperturbed state, the cell populations within the IIRABM remain essentially stable over time. Each cell type has its behavioral rules specified in a defined code block for that specific cell; the sequence of receptor activation is coded in reverse order to avoid the immediate activation of signaling pathways as noted above. Perturbations can occur either via the application of tissue trauma or infectious agents, applied in a generally circular fashion to simulate localized initial insult. The degree of the initial insult can be controlled, and simulation experiments that sweep over values of the initial insult are a primary means of assessing the behavior of the IIRABM. Interested parties are encouraged to download the IIRABM from the website noted above and look directly at the code.

2.8 Parameters, Tuning and Plausibility Testing (See Note 3)

Once the general structure of the rule systems has been extracted from the reference texts and converted into computer code, a series of numerical constants must be added to translate those values. These constants are called parameters and are most easily thought of in terms of a mathematical equation where they are the constants that modify the variables. For example, in the simple equation $ax = y$, "a" is the constant (parameter) that affects the variable "x." In an ABM, the variables are the entities included in the model (i.e., mediators, receptors, cell population counts) and parameters are the constants necessary to calculate those values. The importance of parameters is readily apparent, as they, to a great degree define how the system behaves. There is an even deeper implication of parameters for biological ABMs that will be addressed in the next section, but for the moment we will describe the classical way that parameters are dealt with in computational modeling. As noted above, once rules are converted into computer code some values for these parameters need to be included to run the simulation. If specific parameters are known from the literature (such as kinetic rate constants for biochemical reactions or life-spans for cells) these can be inputted into the ABM, though care must be taken to reconcile units across the model if specific values are used. However, more commonly, exact values are not known (or may be functionally unknowable), requiring an initial educated "guess." The important realization here is that, in almost all cases, the absolute values of the parameters are less important than the relative relationship between parameters across the model. For instance, when implementing a feedback loop structure, it is the relationship between the parameters in the forward- and negative-feedback rules that govern how that loop behaves. Therefore, the first step of dealing with parameters involves attempting to find some set of relational parameters that allows the ABM to produce qualitatively plausible behavior (*see* **Note 3**). At first pass this could be something as "simple" as having the ABM maintain dynamic stability in an unperturbed state (i.e., it does not die off or explode in growth). This is done by stepwise "tuning" of the parameters in the algebraic relationships between the variables. As noted above, since these rules can be considered as computational equivalents to difference equations, changing a parameter is akin to changing the slope of a particular kinetic curve, while changing the algebraic relationship from summation to multiplication will change the kinetic curve from linear to exponential. Tuning is often guided by heuristics and experience. "Hand-fitting" (essentially iteratively trying something, seeing what happens and adjusting) is the most common, and recommended, approach for novice modelers. While laborious, the process allows the modeler to gain insight into how their ABM behaves. Inability to effectively tune a model at this point suggests an intrinsic flaw in the underlying conceptual model. A common challenge to calibration is the lack of sufficient experimental data against which to "fit" the model (not enough reference points exist to refine the ABM to a particular level of confidence). This is unfortunately a very common case in biomedical ABMs, particularly when they are intended to represent clinical scenarios. In these cases, one must fall back upon qualitative interpretation of the ABM's behavior. In this scenario, it is critical to explicitly describe those limitations when conclusions are drawn and communicated.

As an example, the rules in the IIRABM are implemented as either Boolean statements (true/false logical statements) or as arithmetic equations (mediators are summed or divided). Parameters are the values of Boolean thresholds, or degrees by which cytokines, mediators, or receptors increase. As noted above, the absolute values of these parameters are less important than how they relate to each other, and the consequences of these relationships can only be seen via execution of the ABM. Therefore, the overall plausibility of a given set of parameters should be assessed at the population level by running a simulation. These metrics are grossly divided into assessments of microstates and macrostates. Microstate characterizations are essentially trajectories of sums of variables for an individual run (equivalent to an individual animal or patient), and then evaluated over a population of runs. As always, plausible or recognizable behavior is the initial goal. For instance, the IIRABM is able to provide dynamically stable inflammatory cell populations if run in the unperturbed state, and the degree of rise in leukocytes in response to a simulated infection should be matched to the clinical scenario. A more detailed example can be seen in Fig. 3 (from Ref. 13), which depicts the trajectories of total amounts of Tumor Necrosis Factor and Interleukin-10 in a population with severe sepsis. When interpreting this plot, one should recognize that systemic measurements of pro- and anti-inflammatory cytokines in sepsis are notoriously difficult and variable. Clinically, this is heavily influenced by the fact that the starting point of sepsis (called Time-0) is usually impossible to determine. Even for in vivo experiments, since these measurements are often plasma values, there is a discordance between the production of these mediators in the tissues (and their corresponding biologically active tissue levels) and what "spills over" into the blood. However, despite these limitations of clinical and experimental biology, researchers readily construct qualitative patterns of how these mediators are generated and behave. Since the IIRABM does not make a distinction between tissue and circulating levels it is able to reflect the conceptual interpretations of sepsis researchers.

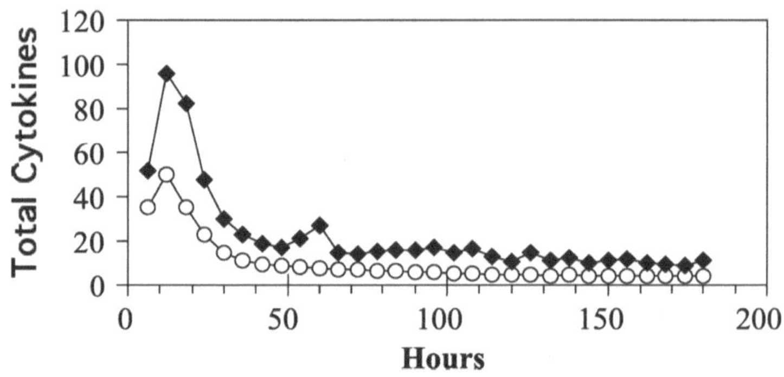

Fig. 3 Patterns of mean cytokine levels (TNF and IL-10) in IIRABM. N = 100 simulation runs with initial conditions including initial infectious insult producing mortality rate of 38%. Note that model is qualitative only and therefore the actual numerical values of the cytokine levels are dimensionless. Note the concurrency in the trajectories of the pro- and anti-inflammatory cytokine levels (not generally recognized in the late 1990s, but by now generally accepted [29, 30]). (Figure reproduced with permission from Ref. 13)

Macrostate criteria for plausible behavior can be seen in population level runs across a range of initial infection levels. Specifically, these simulations will assess whether the IIRABM is a robust enough representation of the key features of the acute systemic inflammatory response such that it can capture the range of behaviors seen in the clinical setting. Four classes of responses to infection are readily apparent: (Class 1) the infection is contained and the system heals ("Healing"; Fig. 4a); (Class 2) the infection overwhelms the inflammatory response, resulting in a progressive increase in infection and system damage, leading to death ("Overwhelming Infection"; Fig. 4d); (Class 3) the infection triggers an intense inflammatory response that is able to contain the infection, but given the intensity of the inflammatory response, it produces enough ongoing damage such that the system dies ("Hyper-Inflammatory Death"; Fig. 4c); and (Class 4) the infection is contained and pro-inflammation contained, but in doing so the anti-inflammatory response is so potent that it renders the system vulnerable to subsequent trivial infections ("Immuno-paralyzed Death"; Fig. 4b). The IIRABM is able to generate each of these four trajectories across the range of initial infectious insults (*see* Fig. 4a–d, reproduced with permission from Ref. 13). The importance of being able to generate these recognizable trajectories cannot be overstated, particularly with respect to Class #4 (Immuno-paralyzed Death), as it is perhaps the most prevalent trajectory seen in modern ICUs [31–34] and yet the most difficult to experimentally reproduce, for obvious logistical reasons.

The behaviors above were initially achieved with hand-tuning of the IIRABM's parameters using a heuristic approach: values were constrained by qualitative, order-of-magnitude knowledge,

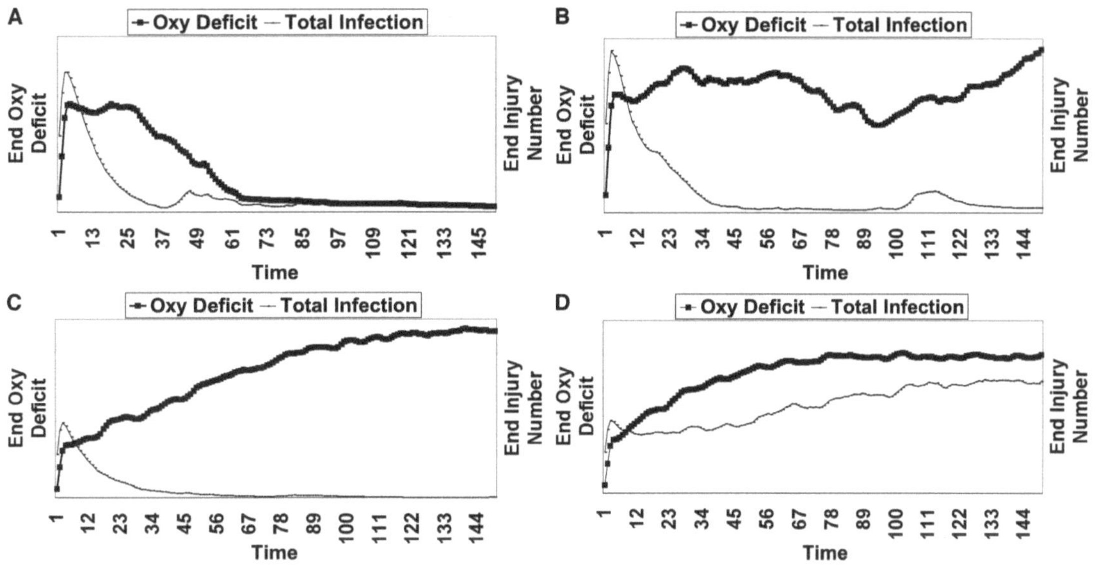

Fig. 4 Representative Individual Runs of the IIRABM demonstrating the four primary trajectories in response to infection. Note the *y*-axis is unitless since the model is qualitative only. (**a**) This figure shows "Healing" with an initial infection number (IIN) = 800. System is able to essentially clear infectious vectors and reduce system damage ("Oxy-Deficit") by around Time 64. (**b**) This figure shows "Immuno-Paralyzed Death" at IIN = 1100. System is able to essentially clear infectious vectors by around Time 40. However, the system is unable to reduce Oxy Deficit sufficiently until a "trivial" second hit around Time 100 is unable to be contained, resulting in system death. (**c**) This figure shows "Hyper-Inflammatory Death" at IIN = 1500. System essentially clears infectious vector by around Time 40, but the forward feedback propagation of proinflammatory pathways leads to progressively increasing Oxy Deficit until system "dies." (**d**) This figure shows "Overwhelming Infection" at IIN = 1800. The system cannot contain the infection, which leads to system death

adjustment, resimulation, and readjustment. This is a standard process in the parameterization of computational models, and usually results in a single or small set of parameter combinations that meet the testing criteria. However, our work in this area has led us to recognize that this is not a sufficient means of representing biology, particularly with respect to the heterogeneity seen in biological behavior. The result of this insight is the subject of the next section.

2.9 Embracing Parameter Space: The Key to Capturing Biological Heterogeneity

"Is it not really strange that human beings are normally deaf to the strongest argument while they are always inclined to overestimate measuring accuracies?"—Albert Einstein [39].

Heterogeneity is both the curse and blessing of biology. A curse in that the intrinsic variability of biological objects bedevils experimentalists attempting to get "clean" data with small error bars. This often devolves interpretation of results into a statistical game of finding the appropriate method that can substantiate "significance." And yet, biology cannot exist without heterogeneity: it is the ability to generate varied responses and trajectories ("phenotypes") from common shared organizational structure that allows biological organisms to survive in variable and shifting conditions, and, more importantly, allows evolution to work. However, the vast majority of modern biological research involves trying to wash-away heterogeneity; the mere goal of creating a reproducible experimental platform is directly opposed to the manifestation of this fundamental biological property [40]. All too often, the result is an experimental platform that is reproducible within an individual lab, but that is incredibly brittle to subtle changes (i.e., changes in reagent suppliers, animal housing practices, lab techs, new investigators, etc.). The other methods presented in this book are examples of an attempt to standardize and codify such experimental practices, but as any experienced researcher knows, the actual implementation of each of these procedures for their own use is a laborious task of tuning, adjustment, and refinement (incidentally, not too different from the process noted above in Subheading 2.8). Therefore, one of the primary goals of translational systems biology and agent-based modeling is to recapture biology's intrinsic heterogeneity, and recognize its critical role (it's a feature, not a bug ...), in investigating biological behavior.

The heterogeneity of biological behavior has two components: a stochastic component (aleatory stochasticity inherent to the system) and heterogeneity that is due to the genetic background and its translation into functional phenotypes. Both properties can be addressed by multiscale mechanism-based computational models, such as ABMs. The aleatory stochasticity is incorporated into stochastic processes in the ABMs, and genetic/phenotypic heterogeneity is treated as a function of the parameter space of the ABM [41]. This latter concept arises from the fact that the various parameters in the ABM determine the impact of any associated equation/rule, and these parameters are thus measures of "responsiveness" of that portion of the model. It is generally accepted that the vast majority of the genotypic variants in biological organisms result in increased or decreased responsiveness of an associated biomolecular pathway or function; this concept maps directly to the role parameters play in a simulation model. A clinical population, then, reflects a distribution of different combinations of parameters for a given ABM, and affects what constitutes a "clinically plausible" representation by that ABM [27].

However, as noted above at the end of Subheading 2.8, most traditional approaches to using dynamic computational models in biomedical research emphasize achieving a high degree of fidelity to experimental data. These data are often generated in the highly constrained biological models that aim to produce "clean," less noisy data. We recognize that this approach vastly limits the generalizable explanatory power of the computational models [41]. Rather than focusing on parameter fitting to find some set that produces a narrowly defined output criteria, we aim to broaden the concept of "fitting" a model to instead characterize its *bioplausible parameter space*. This perspective views intra-population heterogeneity to be the functional result of different levels of responsiveness or "gain" of key cellular/pathway processes that result from genetic or epigenetic diversity. This gain/responsiveness diversity is reflected in the parameters of a computational model, where those parameters affect the impact of their associated rules/equations. Therefore, different parameter combinations reflect the differences of biological responsiveness (and resulting heterogeneous behavior) seen in a population. However, since there is often no direct mapping between the parameters in the model and the genetic/epigenetic factors in an organism/patient, there is no way to presuppose the distribution of viable parameter sets within a population. As such, when we have both simulated disease processes [27] or attempted to discover control strategies [11, 12], we have performed those investigations in a fashion that accounts for the entire bioplausible parameter space in evaluating the generalizability of knowledge produced regarding the dynamics of the systems or the robustness of a potential treatment/control strategy. This approach, however, leads to an exponential increase in the scale of the computational task. In order to make this scale of simulation experiments tractable, we have adopted sophisticated model exploration methods that employ advanced computational methods such as machine learning, active learning and genetic algorithms to "intelligently" explore and characterize model parameter space [42]. While an extensive discussion of these methods is beyond the scope of this chapter, interested readers are directed to Refs. 43, 44 for more technical details.

As an example, heterogeneity in the sepsis population is inherent given the gulf between the various consensus "definitions" of sepsis and the multiplicity of states and trajectories possible for that acute inflammatory response. All too often this heterogeneity is invoked as an explanation for the failure of clinical trials of interventions to treat sepsis, but without any tractable path forward [2, 4, 5]. Based on our insight regarding the role of mechanism-based multiscale computational models as a means of providing a solution to the problem of biological heterogeneity [40], we have used the IIRABM to reconsider the "problem" of heterogeneity in sepsis populations as a function of parameter space [27]. This approach has generated fundamental insights into the nature of sepsis, including: the complex nature of the distribution of sepsis patients (i.e., non-Gaussian) when examined from the perspective of mediator or -omic profiles, identifying the futility of the search for biomarker panels to predict sepsis outcome, and the boundary conditions for putative control strategies for sepsis [27]. The finding from this work suggests that clinical data sets will perpetually undersample the possible biomarker configurations seen in sepsis, and that trajectory analysis is vital for rationally predicting and eventually treating and controlling the disease [40, 45]. Building on our prior work concerning the role of ABMs as a means of developing control/therapeutic strategies [28], we have utilized genetic algorithms/evolutionary computing to define the scope of the task of controlling sepsis [11]. This work aims to answer the question: what would it take to actually control sepsis? Our results suggest that treating sepsis would require tuning multiple mediators in variable configurations at different timepoints, confirm suspicions that researchers may have harbored for years: controlling sepsis is a complex task that involves a series of intricate and variable adjustments at the cellular and molecular level. However, the inability to actually implement such an approach, absent guidance through dynamic computational models, has led to a persistence of attempts to treat sepsis with one or small set of interventions in a one-size-fits-all manner. Our work with the IIRABM proves the futility of this approach, and, more importantly, provides an actual path forward by overcoming the limits imposed by a failure of imagination.

The next step in the use of ABMs to discover control strategies for sepsis draws upon recent advances in artificial intelligence (AI), specifically the use of model-based deep reinforcement learning (DRL). DRL is a type of machine learning that is associated with training artificial intelligence systems [46], and was utilized in the highly publicized success of the game-playing AIs from Google's Deep Mind project, AlphaGo, and AlphaZero [47, 48]. DRL trains a controlling AI agent by "playing" a very large number (up to billions) of simulated games; the AI notes the configuration of the board (observation space) and can make a specific series of moves (action space). It remembers the outcome of each particular move in each particular game, and through the experience of playing all these games, is able to derive a decision tree on what the optimal move would be given a specific configuration of the board. The resulting decision tree is the *policy* the controller AI then uses to interact with any subsequent games it plays. The same approach was applied to the IIRABM, treating it as a "game" to be won; the controlling agent observes the state of the IIRABM at each step, and can interact/apply control by manipulating any of 12 cytokines either up or down given a specified range [12].

The model-based DRL performed in our work is currently unique: no prior application of DRL had been applied to a simulation in the biomedical arena. Notably, this is in distinct contrast to a recent publication on the use of reinforcement learning to train an artificial intelligence agent to optimize the current set of interventions for sepsis [49]. This paper is limited by existing data sets of sepsis patients (and, as such, existing therapeutic modalities) and is unable to guide the development of future interventions, that is, new drugs and/or drug combinations. Conversely, the DRL work with the IIRABM mimics the Deep Mind projects in that it is simulation-based and the trained AI allows for the discovery of novel control strategies (i.e., new drugs, drug combinations, and the timing thereof) based on the much broader control/action space. For the initial proof of concept study, the boundaries of the observation and action space of the training were set into the speculative region in order to determine if a robust control policy was even possible: the DRL AI had complete knowledge of the state of the IIRABM as every time step and had the ability to manipulate any of the included mediators within a range of up or down. The result of this work resulted in a trained AI that reduced the mortality of a simulated population from 46% to 0.8%, tested over a set of 500 different parameter set/initial condition combinations with baseline mortality rates in the range 1–99% for each set: these parameterizations spanned the entire space of clinically plausible parameterizations for the IIRABM [37]. Moreover, no parameter set/initial condition combination exhibited an increase in mortality rate compared to baseline; that is, no cohort was detrimentally affected by the learned policy. These results suggest that despite being trained on a single patient parameterization, the policy generalizes well, as it is robust to variations in parameter set/initial condition combinations. It is worth noting that the consistent learned policy does not mean that there was a consistent applied therapeutic regimen: the learned policy provides for an optional control action given a particular state of the IIRABM, whenever that state happens to occur. As a result, the sequence of optimal actions varies based on the specific dynamics of a particular run. This adaptive nature of the control was evaluated by evaluating the dynamic patterns of the applied control between different simulation runs, which demonstrated a range of different patterns that all resulted in successful outcomes.

3 Conclusion

While the area of sepsis research has embraced computational methods related to data analysis, such as network analysis and machine learning, there has been less acceptance of the role of dynamic computational modeling, such as agent-based modeling. This may be largely because the process of developing and using models like ABMs require a much higher degree of active involvement on the part of the sepsis researcher than data-centric approaches. It is often not possible to identify a "hand-off" point where the computer people work their magic and come back with a result. Agent-based modeling and other types of quasi-mechanistic dynamic modeling are ***knowledge-based***, rather than data-centric, and directly involve the conversion of the sepsis researcher's own knowledge and hypotheses about how sepsis works into a computational form. Added to this is the common misperception that

ABMs are not "real." This of course presupposes that the output of highly engineered biological models or sparsely sampled clinical populations are any more "real" in their ability to capture generalizable aspects of the processes that drive sepsis; we have made the argument that they are fundamentally not [40]. Rather, we make the case that the use of dynamic computational models is more in keeping with how other scientific disciplines (e.g., the quantitative sciences such as physics and chemistry), and engineering tasks are addressed: by incorporating abstracted yet generalizable mathematical and computational models that drive investigation, discovery and the generation of solutions. We pose that the recurrent nature of translational failure in the attempts to develop therapeutics for sepsis bespeaks a deeper issue that will not be addressed by finding the "next molecule/mediator/cell type/receptor" that will somehow provide the cure, where the several thousand prior candidates did not. Almost certainly any viable mediator-targeting strategy will involve combinations of drugs, personalized to individual patients; yet the combinatorial possibilities quickly overwhelm any attempt to extract "precision" from clinical data. We contend that the use of multiscale mechanistic models, such as ABMs, are the *only way* to rationally design such multimodal, precise, and personalized therapies [45]. However, we fully recognize that such modeling cannot function without all the existing lines of biomedical investigation. Basic science is essential in acquiring the mechanistic knowledge that informs the models and directs the development of therapeutic compounds. Furthermore, clinical studies are the essential final arena where any proposed intervention strategy must be tested and proven to work. Moving forward and making progress in the future will require an integration of all these types of research into a cohesive multidisciplinary program if the challenge of sepsis is to finally be addressed.

4 Notes

1. The strengths of ABM as a modeling method have been emphasized in the preceding text: intuitive structure facilitating knowledge translation and representation, intrinsic management of spatial issues, the ability to capture complex behavior and similarity in behavior and output to traditional wet lab experiments. These benefits would seem to suggest a fairly generous application of ABM in the biomedical arena. However, there are significant limitations to ABM, as with all modeling methods. These include high computational requirements for large scale models, inability to "formally" analyze the inner workings of the model, difficulty in calibration due to the nonlinear relationships between agent rules and behavior, and difficulty in matching a specific run of a model's

evolving conditions with a real-world reference (such as the case of attempting the predict the outcome of a specific patient). In the discussion of when one should use ABM, it may be more useful to determine those instances where ABM does not suit the modeling problem at hand and the limitations listed above factor into that determination. These instances can mostly be expressed in terms of the suitability of using equation based modeling, which remains the default method of mathematical modeling dynamic system behavior:

(a) Ordinary differential equation (ODE) modeling is preferable if the system can be characterized by well-mixed compartments/populations. In these situations mean field approximations will hold and mass action kinetics approaches can be utilized.

(b) Equation based modeling is preferable if it appears possible to derive "formal" insights into the system's behavior. Systems whose behavior can be characterized with relatively simple order ordinary differential equations may be analyzed mathematically, leading to more comprehensive and general understanding of their dynamics.

(c) Equation based modeling is preferable if the development and calibration reference data already exists in equation form. This is particularly true when calibration is to be performed using an optimization algorithm. Again, in this situation the nonlinearities between rule modification and model behavior make this extremely challenging.

(d) Equation based modeling is preferable when the number of agents needed to be modeled is extremely high. The computational demands of ABM when the number of agents reaches toward the millions and billions are prohibitive. In general, equation based modeling remains the initial approach to modeling dynamic systems (including biological ones) and is well suited to modeling processes such as molecular kinetics, biochemical reactions, and gross physiologic behavior. However, if the criteria mentioned above do not hold, ABM can offer an advantageous modeling approach.

2. Most of the construction pitfalls have been mentioned above in Subheading 2.1 (identifying the purpose of the ABM), Subheading 2.2 (linking the ABM to its reference experimental/ conceptual model), Subheading 2.5 (agent level/class selection) and Subheading 2.6 (agent rule determination). When trying to address these challenges, it is recommended to keep two goals in mind:

(a) Minimize assumptions. It is important to remember the old computer programming adage: "Garbage in, garbage out." Agent rules should be tied as closely as possible to causal mechanisms defined by experiment. When generating state or influence diagrams, each individual step should be identified, and if there is not sufficient data or an intermediate step is unknown, this should be explicitly noted (see below). Recognize that each assumption made raises the risk of petito principii.

(b) Be explicit, both in defining rules, their underlying justification and assumptions involved. Not only does this avoid the programming issues noted above, but it also rigorously forces an objective assessment of a particular conceptual model. Many researchers are unpleasantly surprised when concepts that taken as "given" are dissected in this fashion. This process is, of itself, beneficial in determining where the next set of wet lab experiments may need to be done. But it is also critical in being able to assess why a model has been "broken." An explicit representation of the underpinnings of a model provides a guide to where adjustments need to be made for improvement in the next generation. Finally, explicitness in model description is a vital component of being able to effectively communicate a model to other researchers. Transparent explicitness facilitates the evaluation, understanding and eventual acceptance of a particular model in the research community at large.

3. Perhaps the greatest danger in interpreting the behavior of an ABM (or any model, for that matter), is assuming that the model represents some sort of objective truth. In other words, a model that appears to match the behavior seen in the real world is possibly only one of many *plausible* models that fit the data. Therein lies the key to using ABM as a test of the verity of a conceptual model: only a negative result can provide definitive information, that is, the conceptual model is incorrect, were as a positive result can only suggest that the conceptual model *may* be correct. The goal of model interpretation, then, is to develop models that, when broken, can provide some insight as to why they are broken, and the process of model utilization is to sequentially generate models, break them through falsification, and use that information to generate the next model.

4. It is recommended to use existing general purpose ABM toolkits for at least the initial attempts in constructing a biomedical ABM. Even if the researcher/modeler is an experienced computer programmer and wants to write a special purpose ABM, developing a familiarity with existing methods of implementing an ABM will aid in addressing some specific programming challenges involved (i.e., object class definition, emulated

parallel actions, spatial topography development, etc.). One of those toolkits, Netlogo [38], was used to create the IIRABM. Netlogo was originally designed to teach primary and secondary school students the dynamics of complex systems such as bird flocking, fish schooling, traffic and ant colony behavior, and as such is very amenable to novices to computer programming. It has subsequently evolved into a very powerful modeling environment, particularly for developing the qualitative/semiquantitative knowledge translation models described in this chapter. Netlogo is freely available for download at http://ccl.northwestern.edu/netlogo/, and is available for Windows, Macintosh, and Linux. The IIRABM itself is available for download at: http://ccl.northwestern.edu/netlogo/models/community/Innate%20Immune%20Response. Netlogo is also closely related to another introductory ABM toolkit called Starlogo, which is available for download at http://education.mit.edu/starlogo/. Individuals with a background in computer programming might be interested in the following open-source ABM toolkits.

RePast https://repast.github.io/

Mason https://cs.gmu.edu/~eclab/projects/mason/

Agents.jl https://juliadynamics.github.io/Agents.jl/stable/

Finally, ComSES https://www.comses.net/ is a community resource for those interested in computational modeling, particularly agent-based modeling.

References

1. Angus DC, Linde-Zwirble WT, Lidicker J et al (2001) Epidemiology of severe sepsis in the United States: analysis of incidence, outcome, and associated costs of care. Crit Care Med 29:1303–1310
2. Angus DC (2011) The search for effective therapy for sepsis: back to the drawing board? JAMA 306:2614–2615
3. Buchman TG, Billiar TR, Elster E et al (2016) Precision medicine for critical illness and injury. Crit Care Med 44:1635–1638
4. Cross AS, Opal SM, Bhattacharjee AK et al (1999) Immunotherapy of sepsis: flawed concept or faulty implementation? Vaccine 17: S13–S21
5. Cross AS, Opal SM (2003) A new paradigm for the treatment of sepsis: is it time to consider combination therapy? Ann Intern Med 138:502–505
6. Vodovotz Y, An G (2014) Translational systems biology: concepts and practice for the future of biomedical research. Elsevier, Amsterdam
7. Vodovotz Y, Csete M, Bartels J et al (2008) Translational systems biology of inflammation. PLoSComputBiol 4:1–6
8. An G, Faeder J, Vodovotz Y (2008) Translational systems biology: Introduction of an engineering approach to the pathophysiology of the burn patient. J Burn Care Res 29:277–285
9. An G (2004) In silico experiments of existing and hypothetical cytokine-directed clinical trials using agent-based modeling. Crit Care Med 32:2050–2060
10. Clermont G, Bartels J, Kumar R et al (2004) In silico design of clinical trials: a method coming of age. Crit Care Med 32:2061–2070
11. Cockrell RC, An G (2018) Examining the controllability of sepsis using genetic algorithms on an agent-based model of systemic inflammation. PLoS Comput Biol 14:e1005876

12. Petersen BK, Yang J, Grathwohl WS et al (2019) Deep reinforcement learning and simulation as a path toward precision medicine. J Comput Biol 26(6):597–604
13. An G (2004) In silico experiments of existing and hypothetical cytokine-directed clinical trials using agent-based modeling. Crit Care Med 32:2050–2060
14. An G (2006) Concepts for developing a collaborative in silico model of the acute inflammatory response using agent-based modeling. J Crit Care 21:105–110; discussion 110-1
15. Zhang L, Athale CA, Deisboeck TS (2007) Development of a three-dimensional multiscale agent-based tumor model: simulating gene-protein interaction profiles, cell phenotypes and multicellular patterns in brain cancer. J Theor Biol 244:96–107
16. Bonabeau E (2002) Agent-based modeling: methods and techniques for simulating human systems. Proc Natl Acad Sci U S A 99 (Suppl 3):7280–7287
17. Bankes SC (2002) Agent-based modeling: a revolution? Proc Natl Acad Sci U S A 99 (Suppl 3):7199–7200
18. Schnell MA, Zhang Y, Tazelaar J et al (2001) Activation of innate immunity in nonhuman primates following intraportal administration of adenoviral vectors. Mol Ther 3:708–722
19. Thorne BC, Bailey AM, Peirce SM (2007) Combining experiments with multi-cell agent-based modeling to study biological tissue patterning. Brief Bioinform 8:245–257
20. Mansury Y, Diggory M, Deisboeck TS (2006) Evolutionary game theory in an agent-based brain tumor model: exploring the 'Genotype-Phenotype' link. J Theor Biol 238:146–156
21. Bailey AM, Thorne BC, Peirce SM (2007) Multi-cell agent-based simulation of the microvasculature to study the dynamics of circulating inflammatory cell trafficking. Ann Biomed Eng 35:916–936
22. Tang J, Ley KF, Hunt CA (2007) Dynamics of in silico leukocyte rolling, activation, and adhesion. BMC Syst Biol 1:14
23. Walker DC, Hill G, Wood SM et al (2004) Agent-based computational modeling of wounded epithelial cell monolayers. IEEE Trans Nanobioscience 3:153–163
24. Pogson M, Smallwood R, Qwarnstrom E et al (2006) Formal agent-based modelling of intracellular chemical interactions. Biosystems 85:37–45
25. Broderick G, Ru'aini M, Chan E et al (2005) A life-like virtual cell membrane using discrete automata. In Silico Biol 5:163–178
26. Ridgway D, Broderick G, Ellison MJ (2006) Accommodating space, time and randomness in network simulation. Curr Opin Biotechnol 17:493–498
27. Cockrell C, An G (2017) Sepsis reconsidered: identifying novel metrics for behavioral landscape characterization with a high-performance computing implementation of an agent-based model. J Theor Biol 430:157–168
28. An G, Fitzpatrick BG, Christley S et al (2017) Optimization and control of agent-based models in biology: a perspective. Bull Math Biol 79:63–87
29. Tamayo E, Fernandez A, Almansa R et al (2011) Pro- and anti-inflammatory responses are regulated simultaneously from the first moments of septic shock. Eur Cytokine Netw 22:82–87
30. Osuchowski MF, Welch K, Siddiqui J et al (2006) Circulating cytokine/inhibitor profiles reshape the understanding of the SIRS/CARS continuum in sepsis and predict mortality. J Immunol 177:1967–1974
31. Hotchkiss RS, Monneret G, Payen D (2013) Immunosuppression in sepsis: a novel understanding of the disorder and a new therapeutic approach. Lancet Infect Dis 13:260–268
32. Hotchkiss RS, Monneret G, Payen D (2013) Sepsis-induced immunosuppression: from cellular dysfunctions to immunotherapy. Nat Rev Immunol 13:862–874
33. Boomer JS, To K, Chang KC et al (2011) Immunosuppression in patients who die of sepsis and multiple organ failure. JAMA 306:2594–2605
34. Ferguson N, Galley H, Webster N (1999) T helper cell subset ratios in patients with severe sepsis. Intensive Care Med 25:106–109
35. An G (2008) Introduction of an agent-based multi-scale modular architecture for dynamic knowledge representation of acute inflammation. Theor Biol Med Model 5:11
36. Seal JB, Alverdy JC, Zaborina O et al (2011) Agent-based dynamic knowledge representation of *Pseudomonas aeruginosa* virulence activation in the stressed gut: towards characterizing host-pathogen interactions in gut-derived sepsis. Theor Biol Med Model 8:33
37. Cockrell C,An G (2017) Sepsis Reconsidered: Identifying Novel Metrics For Behavioral Landscape Characterization With A High-Performance Computing Implementation Of An Agent-Based Model. bioRxiv:141804
38. Wilensky U (1999) NetLogo. http://ccl.northwestern.edu/netlogo

39. Einstein A, Born M (2005) Letter to max born. In: The born-Einstein letters: 1916-1955. Palgrave Macmillian, London, p 238
40. An G (2018) The crisis of reproducibility, the denominator problem and the scientific role of multi-scale modeling. Bull Math Biol 80:3071–3080
41. An G (2018) The crisis of reproducibility, the denominator problem and the scientific role of multi-scale modeling. Bull Math Biol 80:3071–3080
42. Ozik J, Collier NT, Wozniak JM et al (2016) From desktop to large-scale model exploration with Swift/T. In: 2016 Winter Simulation Conference (WSC) 2016. IEEE, Piscataway, New Jersey
43. Cockrell C, Ozik J, Collier N et al (2019) Nested Active Learning for Efficient Model Contextualization and Parameterization. bioRxiv:644401
44. Cockrell RC,An G (2019) Genetic Algorithms for model refinement and rule discovery in a high-dimensional agent-based model of inflammation. bioRxiv:790394
45. An G, Day J (2019) Precision systems medicine: a control discovery problem. In: Reference Module in Biomedical Sciences. Elsevier, Amsterdam
46. Lillicrap TP, Hunt JJ, Pritzel A et al (2015) Continuous control with deep reinforcement learning arXiv preprint arXiv:1509.02971
47. Silver D, Huang A, Maddison CJ et al (2016) Mastering the game of go with deep neural networks and tree. Search 529:484
48. Silver D, Hubert T, Schrittwieser J et al (2018) A general reinforcement learning algorithm that masters chess, shogi, and Go through self-play. Science 362:1140–1144
49. Komorowski M, Celi LA, Badawi O et al (2018) The artificial intelligence clinician learns optimal treatment strategies for sepsis in intensive care. Nat Med 24:1716–1720

INDEX

Wendy E. Walker (ed.), *Sepsis: Methods and Protocols*, Methods in Molecular Biology, vol. 2321,
https://doi.org/10.1007/978-1-0716-1488-4, © Springer Science+Business Media, LLC, part of Springer Nature 2021